WILD WOMEN FOR GOOD

KATHIE AND ED COX JR. BOOKS ON CONSERVATION LEADERSHIP

Sponsored by

THE MEADOWS CENTER
FOR WATER AND THE ENVIRONMENT
TEXAS STATE UNIVERSITY

Andrew Sansom, General Editor

PIONEERING WOMEN: LEADERS AND TRAILBLAZERS

TEXAS WOMAN'S UNIVERSITY
JANE NELSON INSTITUTE
for WOMEN'S LEADERSHIP

Sponsored by Nancy and Ted Paup

TEXAS WOMAN'S UNIVERSITY

AnaLouise Keating, General Editor

With assistance from Kimberly C. Merenda

Additional support for this book was graciously provided by Will and Pamela Nelson Harte.

Wild Women for Good

Stories of Conservation in Texas

Jennifer L. Bristol

Texas A&M University Press
College Station

First edition
∞ This paper meets the requirements of ANSI/NISO Z39.48-1992
(Permanence of Paper).
Binding materials have been chosen for durability.
Manufactured in the United States of America

Library of Congress Cataloging-in-Publication Data

Names: Bristol, Jennifer L., 1971– author
Title: Wild women for good: stories of conservation in Texas / Jennifer L. Bristol.
Other titles: Kathie and Ed Cox Jr. books on conservation leadership
Description: First edition. | College Station: Texas A&M University Press, [2026] |
Series: Pioneering women: leaders and trailblazers | Includes index.
Identifiers: LCCN 2025020580 (print) | LCCN 2025020581 (ebook) |
ISBN 9781648432606 hardcover | ISBN 9781648432613 ebook
Subjects: LCSH: Women conservationists—Texas—Biography |
Conservationists—Texas—Biography | Women in conservation of natural
Resources—Texas—History | Conservation leadership—Texas—History |
Biodiversity conservation—Texas—History | Wildlife
Conservation—Texas—History | LCGFT: Biographies
Classification: LCC QH76.5.T4 B75 2026 (print) | LCC QH76.5.T4 (ebook) |
DDC 333.7209764—dc23/eng/20250922
LC record available at https://lccn.loc.gov/2025020580
LC ebook record available at https://lccn.loc.gov/2025020581

Published by Texas A&M University Press
John H. Lindsey Building, Lewis Street
College Station, TX 77843
www.tamupress.com

Unless otherwise indicated, photos are taken by the author.

Title page photo courtesy of Sisters of Charity of the Incarnate Word Archives, San Antonio, Texas.

Dedicated to

The ladies of WOE and their daughters

And to

Every woman endeavoring to conserve and protect our natural world

CONTENTS

FOREWORD

Of all the many natural resource books published through the years by the partnership between Texas A&M University Press and the Meadows Center for Water and the Environment at Texas State University, Jennifer Bristol's *Wild Women for Good* is surely among the most inspiring. From the women who formed the first Audubon chapters to stop the plume trade to Rachel Carson's epic *Silent Spring*, which alerted us to the reality that we could pollute the whole world, women have consistently and without a lot of recognition played key roles in the conservation movement. Here on these pages Bristol eloquently describes those early Audubon days and, more recently, the key role Texas women played in the creation of the Texas state park system.

For me, it's personal, as I learned to fish from my grandmother, who lived on a farm in Alabama, and watched as a child as my mother's love of nature was manifest in her remarkable green thumb. After college, as a junior aide at the Interior Department in Washington, DC, I was introduced to the incredible biodiversity of the Big Thicket by Maxine Johnston and Geraldine Watson, who, through years of conflict and inaction, refused to give up on the creation of a national biological preserve. Later, as a budding conservation professional in Texas, I traveled throughout the state meeting with garden clubs all over Texas, without which my employer, the Texas Nature Conservancy, would not have survived those early days.

During my years at the Texas Parks and Wildlife Department, I had the honor of working with remarkable women like Carolyn Vogel, who championed protection of the San Jacinto Monument and Battleground and who, along with the amazing Terry Hershey, created the Texas Land Trust Council, which now represents over forty land trusts across the state. During those days, I worked alongside Lydia Saldaña, who may well be the leading conservation communications professional in the United States today.

It was later my privilege to work with Audubon Texas to create the Texas Women in Conservation Awards in Terry Hershey's name and to honor such female conservation heroes as Blair Fitzsimons, Karen Hixon, Carol Dinkins, and the incredible Virginia Rose.

Here in *Wild Women for Good*, Jennifer Bristol, herself a Texas conservation heroine, tells the stories of these and many more women who have helped make Texas a better place for all of us and for the wild creatures who came before us. Bristol was supported in this work by a partnership with Texas Woman's University and by the Jacob and Terese Hershey Foundation and the Texas Natural Resources Publication Endowment, founded by Kathie and Ed Cox Jr. and Will and Pam Nelson Harte.

For me, perhaps the greatest inspiration today is to watch my own daughter, April, who is now an emerging conservation leader among this generation of wild women for good. I know that many of the leaders Bristol introduces us to on the following pages had a major influence on April, and for that and all the other good things they have done for the rest of us, I am deeply grateful.

ANDREW SANSOM

PREFACE

I was hiking at Bastrop State Park, where I was a ranger, with my mother, Valarie S. Bristol, when she first mentioned she wanted someone (not me at the time) to write a book about women's contributions to conservation. The thought stuck in both our minds, and we revisited it over the years, usually while on a birding road trip.

My mother spent most of her career focused on conservation. We will get to that in a later chapter. As for me, my conservation journey started at the tender age of six. My mother would walk with me as I rode my pony deep into the heart of the Wild Basin Preserve near our home just west of Austin, Texas. We would picnic along Bee Creek, where I fell in love with the rugged hills, wildlife, and sense of freedom I found while riding my sure-footed pony. When I reflect on that time, I can remember every element. The sound of the creek spilling over the rocks. The smell of Ashe juniper and oak forests. The rush of exhilaration as I clung to my pony's neck while she galloped up the big hill.

That deep childhood love was challenged when road crews started blasting their way through that big hill as they built the Loop 360 Highway. The sense of loss was profound. In 1977, that love compelled me, and I had my first foray into fundraising. I joined the cause to "Save the Wild Basin." My mother allowed me to join other volunteers to sit outside Rylander's Grocery Store in West Lake Hills to talk about the project and ask for donations. I decorated a special coffee can in art class with paint, yarn, and feathers. As people entered or exited the store, I would shake my coffee tin and ask for a donation. I have no idea how much we raised, but I remember feeling good about the experience. It was the first time I raised money for conservation, and it certainly would not be the last.

Jennifer L. Bristol on her pony ready to explore the Wild Basin from the stable at the corner of Loop 360 and Bee Caves Road, Austin, Texas. 1977. Courtesy Jennifer L. Bristol

Like many women in this book, I saw my own career take many twists and turns before I was able to align it with my true passion for nature.

In 2015, I felt the nudge again to write this book at the first Audubon Texas Women in Conservation Terry Hershey Awards banquet in Houston. That same year, my mother was battling breast cancer. A few years later, in 2018, she was honored with the Terry Hershey Award and the door opened a little more. That same year, I took a daring step to author my first book, *Parking Lot Birding: A Fun Guide to Discovering Birds in Texas*. In that process I learned what it took to write and publish a book.

The year 2019 was pivotal for me as I stood at the crossroads of my career with the Texas Parks and Wildlife Department as the director for the Texas Children in Nature Network. I was standing there looking carefully both ways and considering my options when I had dinner with one of my best friends since kindergarten, Jennifer Sabo Willcott. Jennifer had been very sick that year, and at dinner she said, "What are you waiting for? You need to write, it's who you are, and you need to tell this story. Your mom's not getting any younger." Five days later Jennifer passed away.

It was the shove I needed. A few weeks later I tendered my resignation to my wonderful supervisors at the Texas Parks and Wildlife Department and to the members of the Texas Children in Nature Network. I was so nervous going into the office of the executive director of Texas Parks and Wildlife, Carter Smith, that day to inform him of my choice. However, his exuberance for the project bolstered my faith that this was the right choice. By the end of the conversation, he was encouraging me to "get to it."

And so, Mom and I started our journey. First, we sat down with Shannon Davies, then director of Texas A&M University Press, to inquire if she thought the press would be interested in the project. She laughed and said, "Of course." Second, we sent out emails and called our networks to ask whom they thought should be included, and we were blown away by the responses. Third, I had lunch with Andrew Sansom to ask him for his thoughts, and he responded with the best answer ever when he said, "You need to apply for a grant. And I can help you with that." With his support, and that of David Todd at the Conservation History Association of Texas, we applied for and were awarded a grant from the Jacob and Terese Hershey Foundation.

With those pieces in place, we started conducting interviews and doing the research. Mom took the lead on the research while I worked to transcribe the interviews and contextualize their stories. Inevitably, each time we interviewed a person, they suggested someone interesting to look into. And they all asked if Valarie was going to be included. Not every one of their recommendations made it into the book, but they were all considered.

We were just finding our stride when the COVID-19 pandemic hit and the world stopped. While the pandemic was awful, it was the pause I needed, as the book was lacking an important element—my voice. I knew why I wanted to write the book. I knew how to write it. What I didn't know was in what way I wanted to present each person. As a history major, I have read many books that were bogged down in the minutia or presented things in a flat, unrelatable way. I knew I didn't want to write that kind of book. I also knew I didn't want to write something fluffy either. I kept coming back to one word: *inspire*.

I wanted to tell the story of conservation in a way that inspired others to step up and take action. I wanted readers to see their own story laced throughout the struggles, challenges, delights, and accomplishments of these women. I also wanted to tell the whole story of conservation, from the preservation of a single species, to the establishment of the Big Thicket National Preserve, to securing policies for water conservation. All of it, and everything in between. I guess I can say that COVID-19 helped me find the purpose for the book.

From the beginning we knew we were going to face several challenges. The biggest challenge in both our minds was the inclusion of women of color. We knew we wanted to include as many as we could find—finding them was the hard part. For most of the history of conservation in Texas, it has been the cause of white, socioeconomically advantaged women and men.

Up until recently, women of color were not usually invited to the conservation table. And if they were, their voices were slow to be heard. I even witnessed that occurrence in my own career. While we were able to find several stellar women, we also acknowledge that there are many young women of color who are starting or are in the midst of their conservation journey. They are the future and we certainly need a more inclusive mix of women to continue and improve the efforts of protecting the environment for future generations. It is exciting to see programs such as Latino Outdoors, Outdoor Afro, Black Women Who, and Black AF in STEM gain momentum to connect people with the outdoors and introduce them to careers.

We acknowledge that each woman's story does not happen in a vacuum. Their parents, spouses, friends, coworkers, business partners, children, and fellow volunteers all helped them achieve their goals. In that mix are a whole cast of wonderful men also working to conserve the wild places and wild things of Texas. When and where we could, we have celebrated their participation and contributions.

Many of the accounts in this book come from in-person interviews or newspaper articles. Interviews are wonderful in that the interviewee can share their emotions and describe their actions and achievements.

However, at times it is difficult to accurately capture relevant dates or other small details. We tried our best to verify every element, but on occasion we simply had to take the oral account as the only place that information was shared. Newspapers offer a good account of events; however, they can have facts that are misstated or incorrect. When possible, we tried to cross-reference those details. Since most of the information was not covered in other sources, we were limited. With that said, we hope that we captured each cause, effort, and story as accurately as possible.

Women have been instrumental to conservation since the late 1800s in Texas. However, it was not until they had equality under the law in the areas of banking, access to careers, health insurance, and admission to the political system that they fully realized their potential with respect to conservation (and everything else).

In a state that only has 6 percent public lands and a staggering 94 percent in private ownership, private funds and partnerships are critical to achieving any aspect of conservation. Government investment goes back to the foundation of the Lone Star State. While positive policies continue to evolve, access to funding to purchase large pieces of lands or fund native plant research or build a nature center all take additional private investment.

For over a century, very few women in Texas had access, on their own, to the level of finances it takes to invest in conservation. Today, women are more in control of their financial independence and the causes they support. Women's financial fitness has also allowed more women to enter the political realm, where they can influence projects and policies relating to parks, water, pollution controls, species protection, and land conservation. Women are also finally welcomed into careers such as biology, forestry, and land management that for decades were controlled, unconsciously and openly, by men. As these doors continue to open, more women are willing to walk through.

This book focuses on the women who dared to be first. Those who pushed the boundaries and broke through barriers. Those who balanced the demands of family and the calling of their cause. They ignored social constraints and statements such as, "You don't have a snowball's chance in hell" or "You should be home cooking dinner for your husband." It would be easy to use this book as a venue to belabor the institutionalized constraints placed on women, but I find that path distracting. I acknowledge those issues and barriers, while trying to keep the focus on the inspirations, actions, and outcomes. If those barriers are central to the telling of that person's story, then I include them.

I fully acknowledge that for the first seventy-three years of Texas' existence, women were dancing backward in high heels, as Governor Ann Richards used to say. I will add that they also had one arm tied behind their back without the power to vote or to hold political office. And it would take another forty years for people of color to have the same rights under the law. While women still do not have full equality in the United States, we have come a long way.

I know that this book is only able to scratch the surface of the stories about all the women who contributed to setting aside parks, protecting our rivers and lakes, saving species, or ensuring that future generations have places to play, explore, and connect with nature. While some women, such as former First Ladies Lady Bird Johnson and Laura Bush, have in-depth books written about them, I have tried to be more inclusive to the women whose stories are not well known. Some stories follow the journey of a single person and their achievements, while others share the collaborative efforts of a group of women. I could not include every woman who contributed to the success of a project, but I have tried to focus on the central figures and their actions.

I hope that the sum of these women's efforts tells the story, both known and unknown, of what it takes to protect and preserve the natural and cultural heritage of Texas. But more importantly, I hope that one or all of these women will inspire you to be wild for good.

WILD WOMEN FOR GOOD

CHAPTER ONE

SAVING THE BIRDS

The land that we now know as Texas has been inhabited by humans for approximately 14,000 years. For the first 13,500 years or so, humans lived day to day in balance with the resources of available wildlife, land, and water. That is not to overromanticize the subsistence living and hardship of the first people; it is to illustrate that it was not until humans altered the balance of nature after the arrival of European settlers that the need to protect and conserve the natural world arose.

After the end of the American Civil War, Texas experienced a steady rise in population as families, speculators, entrepreneurs, and war veterans hung out their signs saying "Gone to Texas." Cities, towns, ports, ranches, farms, and forts became part of the new tapestry that blanketed the land. As the human threads wove tighter, the abundance of species began to diminish. In the expanding urban centers, women started to ask: Where would our children play? Would their futures be filled with smokestacks, dust blown from overworked lands, acres of stumps

"The woman behind the gun." Gordon Ross. Illustration shows a woman, possibly Coco Chanel, wearing a large hat with feathers, shooting at large white birds with a rifle; two dogs labeled "French Milliner" place the dead birds on a pile at her feet. New York: Keppler and Schwarzmann, 1911. Library of Congress Prints and Photographs Division, Washington, DC, http://hdl.loc.gov/loc.pnp/pp.print

Betty Neal with her two daughters, Mary (*center*) and Ruby Irene (*right*). Postcard, November 13, 1911. Courtesy Robert Hanks Brister Papers, Special Collections, The University of Texas at Arlington Libraries

where forests once stood, and feathers that could only be found on hats and not birds? Thankfully, their answer was, "Not on our watch."

To ensure that their children had an agreeable future, the women organized in clubs and held luncheons and discussion groups to understand how to conserve their fading natural and cultural heritage. They did not have a strategic plan or a road map, but they knew that, left unchecked, the future ahead of them would be unacceptable for their children. That was enough to spur them to take action. They were not always able to break down the barriers to their success; however, they stood at those gates and, when possible, slipped passed and extended their hands to pull others through.

THE FIRST AUDUBON SOCIETIES IN TEXAS

Cecelia E. Seixas (1883–1900) and Estelle Cannon Hertford (1876–1946)

The Second Industrial Revolution was in full swing by the 1880s in the United States. Steel plants churned out the necessary materials to fuel new skyscrapers or provide the railroads with tracks to extend to the timber country in the Northwest. While factories and cities grew, so too grew pollution and the trappings of urban life. Out of the urban cores, a new class emerged stronger than ever. The middle class consisted of small-business owners, lawyers, doctors, accountants, and other professionals who could earn enough money to no longer be counted as poor but who were not yet rich enough to be part of the wealthy elite.

While the number of middle-class citizens was increasing in the United States and Europe, the strict rules of social conduct of the Victorian era were also increasing. Under these rules, there was little room for error for middle-class women. There was also little room for achieving political influence, a higher education, or financial independence. These restrictive norms left many women longing for more than just being a reliable mother or dutiful daughter. Slowly, women began to gather at luncheon groups to share ideas, listen to intellectual speakers, offer gardening tips, and focus on social issues.

Solvay Process Company's works, Syracuse, New York, between 1890 and 1900. Detroit Publishing Company photograph collection, Library of Congress Prints and Photographs Division, Washington, DC

After the Civil War ended, women's clubs began to spring up across the country, and by

1890, the General Federation of Women's Clubs had been organized. Earlier, Jane Croly, a professional journalist, had formed the Sorosis Club after being denied entry—based on her gender—to a lecture by novelist Charles Dickens. The club's original intent was to bring intellectual enrichment to women through guest speakers, book recommendations, and informal classes. To celebrate the Sorosis Club's twenty years of existence, Jane Croly and others organized a convention in New York City. They brought together leaders from women's clubs around the country, including Texas, to organize the federation.[1]

Galveston, the major port in Texas, was filled with new wealth and women seeking to form clubs for various causes. The rise in the middle class also meant that the women of the time tried to keep up with East Coast fashions, which included the use of feathers in hats and even on dresses. The fashion trend did not go unnoticed by those who would rather see the feathers remain on the birds in the wild.

In December 1893, the *Galveston Daily News* ran an article with the provocative headline "Save the Birds. Give Up Feathers on Your Hat If Nothing Else." The author of the article wrote, "Someone has said if women could only know of the destruction of bird life that their love of finery occasions, they would make it unfashionable to wear the feathers of murdered birds. The bird is shot while trying to protect its young ones, and the young birds are left to starve."[2] The author also reported that arsenic was often used to poison the birds, and the toxic chemical would leach out of the feathers after they had been affixed to the hat and give the wearer headaches and blurry vision.

Ornithologist Frank Chapman conducted a study in New York City about women's hats and what types of bird feathers were displayed. The five hundred–plus hats he observed had more than forty species of birds represented. He went on to note that the demand for feathers for the plume trade was so high that in 1886, an ounce of desirable feathers sold for thirty-two dollars, nearly twice the price of an ounce of gold. By 1900 that price had gone up to eighty dollars an ounce and the industry employed thousands of people, making it big business in America and Europe.[3]

The devastation piqued the interest of many Galveston residents, including two young women: Cecelia E. Seixas (age 15) and Estelle Cannon Hertford (age 22). The two saw firsthand the loss of colonial wading

Model wearing "Chanticleer" hat of bird feathers, circa 1912. George Grantham Bain Collection, Library of Congress, cph 3b08931

and other shorebirds in the region. They also observed birds' feathers and wings affixed to the fashionable hats of the society ladies of the day, including those most likely worn by their own mothers and aunts. In 1898, Cecelia and Estelle decided to organize the first Audubon Society in Texas to raise awareness and take action. The Galveston Audubon Society was organized on March 2, 1898, and by 1899 it included seventy-five members who met twice a month.[4]

Naturalist and writer George Bird Grinnell (1849–1938) formed the first Audubon Society in 1886 in New York. Grinnell had been a student of Lucy Audubon (1787–1874), wife of naturalist and painter John James Audubon (1785–1851). Grinnell often reflected on her lessons about stewardship and conservation ethics in his articles in *Field and Stream* magazine. He put Lucy's teachings into direct action when he took on the millinery trade.

Cecelia E. Seixas standing in a frozen garden after a winter storm at Thirty-Ninth Street and Avenue P in Galveston, 1899. This is the only known remaining image of Cecelia. Courtesy Rosenberg Library, Galveston, Texas

Knowing his cause needed a grassroots movement, he formed the Audubon Society. He named it thus to honor his teacher and mentor Lucy, and to utilize the cachet of her husband's famous paintings.[5]

The society proved to be very popular, rapidly gaining 17,723 members, of whom 104 were reported to be living in Texas. However, the society became too much of a burden for Grinnell's staff at *Field and Stream* magazine, and the new organization was abandoned in 1889. The movement to protect birds remained important, and several states began to form ornithological societies, humane societies, or bird-focused committees within existing clubs.[6]

It is likely that in 1899, members of the Galveston Audubon Society voiced their concerns and wrote letters to help pass the Lacy Act of 1900, which prohibited the national trade of wildlife, fish, and plants. Sadly, there is not much record of the organization's meetings, as tragedy struck the community. On September 8, 1900, seventeen-year-old Cecelia and between six thousand and twelve thousand other inhabitants of Galveston and nearby coastal communities died in one of the most devastating hurricanes in the country's history.

Estelle Cannon married Joseph Hertford in 1895. The young couple survived the hurricane, but just days after it swept through the island, Estelle gave birth to her first son on September 25, 1900. Since no surviving records of the other seventy-five Audubon members exist, there is no way of determining their fate in 1900.[7] Estelle went on to live a long life and is mentioned many times in the society pages of Galveston, Houston, and the various cities she visited.

Georgia Brisley Schley in feather hat, Galveston, circa 1890. Courtesy Rosenberg Library, Galveston, Texas

In 1901, the National Committee of Audubon Societies was formed to bring synergy to the local groups that were popping up across the country. The chair of that organization was William Dutcher. In 1902, Dutcher reached out to Houston naturalist Henry Philemon Attwater to request that he start a local chapter. Attwater declined, but not before recruiting Adella Penfield of La Porte, Texas. In the spirit of the Galveston Audubon Society, two teenage girls became the officers: Millie Eva Lamb and Hope Terhune.[8] In 1916 the Galveston Bird and Tree Club formed to promote bird watching, bird protection, and tree planting.

With the re-formation of the National Audubon Society in 1905, a greater focus at the national, state, and local levels was placed on causes contributing to the decline of birds. The Lacy Act opened the door for the Weeks-McLean Act of 1913, which restricted the trade of feathers, ended spring hunting, and established hunting seasons. In 1918, the Migratory Bird Treaty Act was enacted, making it a crime to sell or hunt migratory birds, their eggs, or their feathers. The Migratory Bird Act remains under constant attack, even today.[9]

On March 5, 1999, the Texas legislature issued a House concurrent resolution to recognize the important work of the early Texas Audubon Societies and the legacy of these caring and inspiring young women. The resolution also acknowledged the valuable work of Captain Mervyn Bathurst Davis, a leader in the movement who expanded Audubon to fourteen chapters across Texas.[10]

Since 1923, Audubon Texas has partnered with other organizations to manage Green Island and later the Vingt-et-un Islands, which host the largest Reddish Egret and Roseate Spoonbill rookeries in the world. They manage additional barrier islands along the Texas Gulf Coast that are important nesting and wintering sites for a myriad of shorebirds and gulls. Today, Audubon Texas runs four nature centers and hosts statewide programs, including Bird City Texas and the Lights Out campaigns. There are currently twenty chapters in the Lone Star State.

One hundred fifteen years after Cecelia and Estelle organized the first Audubon Society in the state—and one of the first in the country—Audubon Texas established the Texas Women in Conservation Terry Hershey Award. Audubon Texas board chair George Bristol and vice chair Andrew Sansom met with Terry Hershey to establish the organization's recognition award. Assistant director Haily Summerford had already created a program focused on leadership in conservation for young women. Since 2015, at their annual luncheon, they have honored Texas women who have played a role in conservation and recognized aspiring conservationists through the Conservation Leaders Program for Young Women. Many of the women in this book are the recipients of that prestigious award. Regrettably, we are not able to feature them all.

Today, approximately one hundred million bird watchers are active in the United States. More than sixteen million from that group take bird-specific trips, with the Texas coast and Rio Grande Valley ranking among the most visited destinations. According to the Cornell Lab of Ornithology, gear spent on bird-watching (binoculars and spotting scopes) in 2016 surpassed the purchase of hunting rifles and shotguns for the first time, with an estimated $1.8 billion in sales. According to the 2025 State of the Birds report, the national annual economic output generated by birders is $279 billion. In 2025 that comes to an estimated $5.5 billion in Texas.

Audubon and other ornithological societies that support conservation education and advocate on behalf of those that cannot—the birds—perform invaluable work to foster a sense of place and love of nature. The culmination of this focus on our feathered friends can largely be attributed to the two pioneering young women who dared to care in 1898.

NATURALIST, FOLKLORIST, AND QUEEN OF CAUSES

Bessie McCollough Reid (1880–1962)

As a child, Bessie Reid is reported to have had a love of the outdoors and nature, with a special affinity for birds. As an adult, she and her husband, Bruce Reid, lived in Port Arthur, where he worked as a head electrician for Gulf Oil. The couple had one son, John Reid, born in 1903. Thankfully, Bruce supported his wife's passion for exploring and understanding nature and made no demands for her to be a picture-perfect society wife. He welcomed the menagerie of nonhuman guests that shared their home as Bessie rehabilitated various birds and other animals.

Bessie often described herself as a "joiner" and was a member of nearly every local club in the Port Arthur / Beaumont area. If there was not a

Bessie Reid seated in a garden at her home, date unknown. Courtesy Special Collections, Mary and John Gray Library, Lamar University

Ten hunters with rifles posing with their dogs and the game birds they shot, probably in Webster, Dakota Territory, circa 1890. Approximately forty birds are attached to a Chicago, Milwaukee, and Saint Paul Railway car; a pile of dead birds are behind one hunter. Library of Congress Prints and Photographs Division, Washington, DC, 2012646797

club that captured her interest, she would start one. She was involved with the Texas Folklore Society, American Association for the Advancement of Science, American Ornithological Society, Louisiana Horticulture Society, Houston Nature Club, East Texas Nature Club, Audubon White Heron Society, Humane Society, Texas Garden Clubs, and Big Thicket Association (founding member). Perhaps some of her most important club work occurred around 1914 when she served as chair of the Conservation Committee of the Texas Federation of Women's Clubs.[11]

Like so many nature lovers living along the Texas Gulf Coast, she grew concerned with the decimation of the colonial wading bird population along with other migratory and resident birds. Bessie used her position as Conservation Committee chair to influence others to protect the birds and reduce threats to them.

In a November 29, 1914, article by Decca Lamar West in the *Austin American-Statesman*, Bessie called on her fellow Audubon members and federation sisters to support the recently passed Weeks-McLean Act, which was designed to stop the commercial hunting and illegal trade of migratory birds. The law also put an end to what was called "millinery murder."[12]

The Weeks-McLean Act was not well written, which left it open to attempts by the millinery trade and feather exporters to try to defund its enforcement. Farmers were also against the bill, as they often viewed birds as pests to be eliminated rather than as assets providing economic value by feeding on millions of insects every year. For the most part, public support kept the law intact until it was replaced with the Migratory Bird Act in 1918.

In a May 1914 article, Bessie made the case to support the enforcement piece of the law, stating, "The amount of $100,000 asked for by friends of the biological survey of the Department of Agriculture is very small; but without this fund for the payment of wardens to enforce compliance with its provisions, the law becomes practically a dead letter." She asked sportsmen to support the fund to ensure stable game populations. Her article also appealed to parents and grandparents: "Children also have a right to demand that we pass on to them as full a heritage as we have received from former generations."[13]

In her time as chair of the Conservation Committee, Bessie focused on education and how to use nature as a tool to teach children about science, history, art, and life skills. In another November 1914 *Austin American-Statesman* article, she encouraged parents to allow children to go "hunting with miniature cameras" instead of guns. She also suggested that schools have a "bird day" where students could venture outside to learn about birds and their habitats. The education of children was not her only priority; adult education and species preservation seem to have been equal in her mind. In the same article, she noted that farmers could not be expected to make peace with the birds that ate their crops, and declared it was time to establish a system of bird reserves. Bird reserves, she wrote, "were already being created in more than twenty-six states, including New Mexico and Louisiana, and it was high time for Texas to join the cause."[14]

Reading Bessie's articles making the case for conservation in 1914 is similar to reading a modern-day editorial in *Audubon* magazine. She was passionate about the need for preservation lands, better management of forests (as opposed to clear-cut practices) and protected wetlands, and recognition of the harmful impacts of house cats in urban areas. She even noted a reduction in wild turkey populations across the state and the need for season-specific hunting regulations to preserve the hens and broods. As chair of the Conservation Committee of the Texas Federation of Women's Clubs, Bessie was in high demand as a speaker in cities across Texas and the Southwest. She earned a reputation as a respectable ornithologist who was recognized not only by her peers in Texas but by the Smithsonian Institute in Washington, DC.

With the onset of World War I, Bessie's medical training came in handy as she went to work as the state president of the Surgical Dressing League of the Red Cross. Bessie had attended medical school at the University of Texas before her marriage to Bruce and was pleased to use those skills to teach others how to transform worn-out clothing into surgical dressing. Once the United States entered World War I, it became clear there was a serious lack of medical supplies. She traveled across the state training others how to sterilize and prepare the dressing. Her work with the Red Cross elevated her to war-angel status, which served her well in the years to come when she advocated for conservation lands and policies.[15]

Known for her accuracy in identifying the birds found in each season, Bessie regularly sent reports to the Bureau of Biological Survey and to famed ornithologist Harry Church Oberholser. She taught classes to Audubon and other outdoor clubs, instructing members how to band birds and record their information to compile annual studies on migration patterns and nesting habits. She became the leading authority in Texas on the nesting locations for Roseate Spoonbills, whose population had been decimated when their bright pink feathers became popular for ladies' fans and hats.

As early as 1905, Bessie began to document the spoonbills and other nesting birds on the Vingt-et-un (also spelled Vingtune in some articles) Islands off Morgan's Point in Trinity Bay. Fearing hunters would raid the nesting sites, she did not disclose the information until decades later when she showed members of the Houston Outdoor Nature Club the rookeries in hope of having them protected. Her efforts were a success, and the island rookeries were protected during the breeding and nesting seasons by a game warden and volunteers.[16]

In June 1934, Bessie began corresponding with William Tucker of the Texas Game, Fish and Oyster Commission and Robert Allen of the National Audubon Society. The two men requested her assistance with identifying all of the Roseate Spoonbill nesting sites she had documented in Texas, with the hope of protecting at least some of the locations. In her response, she listed seven nesting colonies that ranged between Port Arthur and Padre Island. In her letter, she politely recognized how much assistance she had received, especially from her friend Corrie Herring Hooks:

> July 17, 1934
> Dear Sir,
>
> Next time you are down this way, I will gladly show you the sites of those seven roseate spoonbill colonies, not counting Vingt'un, to which I took members of the Houston Outdoor Nature Club as a preliminary to their getting your protection for this one of two colonies whose locations I have known for thirty years or more.
>
> The groups I found last year I would like to have known as the "Picton-Reid" colonies. That is a well-earned honor for my husband, who gives me more than we can well afford to pay for my investigations, and for Mr. W. M. Picton of Port Arthur and Rockport, who lent me a seagoing boat, a man to run it, a cook and paid the cost of the trip.
>
> The inland Chambers Co. group . . . should be called Mrs. J. L. Hooks. Mrs. Hooks has repeatedly driven me to areas we wished to examine, was with me when we found the meanest group to get at of the lot, and has literally lent me houses, camp houses, as bases from which to start out in Galveston and Chambers Co.[17]

Her letter also reported that, financially, times were tough, and if not for the assistance of her friends and support of her husband, she would not have been able to continue her observations. Bessie supplemented her husband's income by writing articles for local newspapers and was often paid a small honorarium for her lectures. In the same letter, Bessie included a paragraph explaining that she never traveled on observations alone because of her desire to have a witness to what she was identifying.

What she did not mention in the letter was that she never learned to drive, and Corrie Hooks was often her willing chauffeur. In fact, the two best friends were on their way to Waco for Bessie to give a presentation about wildlife to schoolchildren when they were involved in an accident. A December 17, 1933, headline in the *Waco Tribune-Herald* read, "Crash Kills Two Birds: Naturalist Has Wreck En Route to Waco; Snakes Saved." In the article, it was reported that Bessie "was slightly injured, the blue birds, a strange marsh bird, and other birds and animals with her were killed. Two pet snakes she had in her handbag escaped unharmed." Who keeps snakes in their handbag?[18]

In 1948, there was a proposal to widen and deepen the Trinity Canal as it passed from Galveston to the Houston ports. Thanks to Bessie's continued vigilance, and with the help of Audubon Texas, the ship canal was diverted away from the rookery. (Audubon Texas has overseen the management of the island for decades, and in a 2023 interview, Audubon Texas executive director Lisa Gonzalez mentioned the island is a topic of much discussion as climate change raises the sea level and affects coastal nesting birds.)[19]

Bessie also had a passionate love for the rich biodiversity of the woods of East Texas, in particular around the Big Thicket region. Bessie's friend and fellow naturalist Connie Hagar shared that "Mrs. Reid eschewed all club work except conservation—and that she did mainly in the hope of preserving something of the wilds she loved." Connie and her sister Bert were awed by Bessie's knowledge. Bessie and Corrie Hooks took Connie and her sister exploring many times deep into the piney woods around Village Creek. When they were not exploring together, Bessie, Corrie, and Connie corresponded regularly.[20] The three formed the trifecta for supplying accurate and annual bird data to ornithologist Harry Church Oberholser for his 1,108-page book, *The Bird Life of Texas*.

As a founding member of the East Texas Big Thicket Association, Bessie worked tirelessly to protect the forest from overlogging, oil and gas drilling, and development. She documented plants and wildlife to establish baseline studies for future biologists and spoke eloquently on the topic of preservation to nearly any group willing to listen. During the

1930s, she gave several presentations to the Texas Garden Clubs on the concept of preserving the Big Thicket as a national park. The East Texas Big Thicket Association envisioned a four-hundred-thousand-acre park system of forests, bogs, bottomlands, and prairies.[21]

In 1933, Bessie served as chair of the Nature Study, Birds, Flowers and Wildlife Committee of the Texas Garden Clubs. At that point, the Texas Garden Clubs was garnering support for the beautification of highways by planting eighty thousand trees and wildflowers in preparation for the Texas Centennial. In an *Austin American-Statesman* article about plans for the tree planting, author Elizabeth Kendall described Bessie as "one of the best-informed women in Texas on these subjects," referring to birds, flowers, and wildlife.[22]

But not everyone appreciated Bessie's intellect. Some men were unable to see a woman as an authority on any topic outside the boundaries of hearth and home. In 1937, the Texas Academy of Science held its annual conference in Beaumont, with the major theme being the preservation of the Big Thicket. In the meeting's follow-up report, "Ornithology Lacking," the author did his best to discredit Bessie. He wrote, "Although the Dallas Nature Study Club and East Texas Nature Club were represented at the meeting, an amazing ignorance of bird life was displayed during various sessions. Mrs. Bruce Reid of Port Arthur appears to be the final authority of all matters ornithology in this part of the state and Mrs. Reid is extremely unreliable."[23] His misogynistic statements continued for several pages as he bristled at the idea that Bessie and even Corrie Hooks might be anything other than polite housewives.

The author reversed his tone when he praised the male speakers at the conference. The document was mailed to the entire Texas Academy of Science membership and must have been upsetting for Bessie and Corrie to read.

In addition to being an influential naturalist who inspired others to care about nature, Bessie was a talented storyteller and folklorist. She was close friends with famed folklorist J. Frank Dobie (1888–1964) and his wife, Bertha Dobie (1890–1974). Bertha was also a celebrated writer and garden club member who shared her love and passion for nature in her articles and lectures. She wrote frequently for *Nature* magazine, *Garden Digest*, and *Southwest Review*. The Dobies and Reids often traveled together

ON BIRD STUDY OUTING. The late Mrs. Bessie M. Reid, author and naturalist, LEFT, is shown with Mrs. Corrie Hooks of Beaumont. The picture was taken while the two were studying bird life on Paint Creek Ranch near Junction.

Bessie Reid and Corrie Hooks birding at Paint Creek Ranch near Junction, Texas, date and newspaper source unknown. Courtesy Katherine Korbell Brown

across Texas and the Southwest, including frequent trips to Rockport, where they stayed with Connie and Jack Hagar. On more than one occasion, Bessie took the Dobies exploring in the Big Thicket and recruited them to use their influence to engage and inspire members of their circles to conserve the towering timbers and pockets of prairies.[24]

Bessie also collaborated with Florence Stratton to release a 1936 book about Texas-based Native American legends and stories titled *When the Storm God Rides: Tejas and Other Indian Legends*. According to the Center for History and Culture of Southeast Texas and the Upper Gulf Coast, one of the stories Bessie penned, "The Legend of Kisselpoo," inspired the song "Running Bear" by Beaumont disc jockey J. P. Richardson, also known as the Big Bopper.[25]

Always leading by example, Bessie and her husband decided to bequeath their seventy-five-acre property and home in Hardin County along Village Creek to become part of the Big Thicket National Preserve in 1954.[26]

Bessie continued to educate and advocate through her speaking engagements and articles well into her seventies. In 1958, under the leadership of Magnolia Greene, the newly formed Travis Audubon Society requested that Bessie share snippets of her fifty years of bird study in a presentation she called "Humorous Experiences While Birding." She was, of course, driven to Austin by Corrie Hooks. The presentation was one of the last events Bessie attended outside East Texas before she passed away on April 28, 1962.[27]

Bessie Reid's and Corrie Hooks's legacies can often be cast together, and no better example exists than the gift of the Hooks-Reid Bog. In 1972, the bog, which was named for the two women, was willed to The Nature Conservancy by Armand Yramategui (1923–1970), a local naturalist and vocal advocate for conservation in Southeast Texas.[28]

Armand likely first visited the bog on a nature study excursion with the Houston Outdoor Nature Club, of which Bessie and Corrie were members. Armand carried forward the torch of setting aside lands for the protection of wildlife and for shining a light on environmental issues surrounding Houston until his untimely and shocking murder in January 1970. The Hooks-Reid Bog that he gave to honor these pioneering women is now part of the larger Big Thicket Bogs and Pineland Preserve owned by The Nature Conservancy, and it protects many rare plant species and the unique bog ecosystem.[29]

Bessie Reid's legacy can be found in the many young minds she inspired to care about nature, her advocacy for the preservation of birds and wildlife, and the identification of several new species of plants, including the Reid Violet. Through her work with the Texas Federation of Women's

Clubs and other organizations, she laid the foundation for other trailblazing women. By far, her biggest legacy and gift is forever conserved in the Big Thicket National Preserve. Sadly, she did not live to see it formally designated as the first national preserve in 1974. Because she was one of its early advocates and passed away before its formal designation, she is seldom referenced in newspaper articles or books as one of the key campaigners for the preserve. Thankfully, Big Thicket advocate Maxine Johnston of Beaumont acknowledged Bessie's valuable contribution and retained some of her articles in the Lamar University Archives.[30]

BIRDS, BOOKS, AND THE BIG THICKET

Corrie Herring Hooks (1876–1963)

Corrie Herring and her sister were orphaned at a young age when their parents died in quick succession. The sisters were taken in by an aunt and uncle who did not have children of their own. Corrie is described as having boundless energy as well as great beauty. She married Joseph Lamar "Sam" Hooks in 1892 when she was just sixteen years of age. Twice his bride's age, J. L. Hooks came from a well-established East Texas family that had settled near Kountze, Texas, in the late 1840s. The newlyweds jumped headlong into expanding their family and welcomed their first child, Verna Hooks, to the world in 1893, with six more children to follow.[31]

Like so many women of her station in life, Corrie found intellectual stimulation and a sense of purpose in her family and through club work in her community. She was a charter member of the Texas Ornithological Society, East Texas Nature Club, Magnolia Garden Club, Louisiana-Texas Horticulture Club, and Big Thicket Association and was an active member in the Texas Federation of Women's Clubs.[32]

Through her club work, she made lifelong friends, including Bessie Reid and Connie Hagar. Like Bessie, Corrie was drawn to the study of nature, with a particular focus on birds, and flowers through photography. She often traveled with the Magnolia Garden Club to Austin to showcase her East Texas wildflowers in the annual wildflower exhibit at the University of Texas. Additionally, she was a sought-after speaker and often gave presentations to Campfire Girls, Boy Scouts, Girl Scouts, chambers of commerce, and various topically relevant clubs.

In 1907, if a woman was not traveling with her husband or family, one of the only other socially acceptable ways to explore was to travel with friends or with other club members. When her daughters were twelve and fourteen—old enough to care for their younger siblings—Corrie was ready to discover new horizons. In a comical letter shared by the family, daughter Verna Hooks reported, "We are both so tired, that we wish we had someone

Corrie Hooks portrait, date unknown, from family photos.
Courtesy Corrie Herring Hooks's family

hired. We've ironed nearly everything so that you won't have to do anything. Given the house a general clean and we feel awful green and lean. Brother and Joe are so *bad* and we feel awful sad and wished we'd stayed with Aunt Sally so she'd make them behave her. They are both unmanageable, we wish we were able to make them mind us—but they always fuss."[33]

The sweet, poetic letter also reported that Verna's brother continued to wear the same dirty shirt for days. While the letter might seem like a cry for help from a fed-up teenage daughter and an appeal for a mother's return, it also carries the voice of a strong, independent-minded girl standing on the precipice of adulthood and making light of her situation.

Corrie Hooks holding a horse while collecting honeysuckle and dogwood flowers with her children, March 4, 1911, from family photos. Courtesy Corrie Herring Hooks's family

Corrie clearly trusted her children to care for themselves and continued to travel with and without them until the last years of her long life.

During Corrie's lifetime, newspapers often printed short reports about people's coming and goings. The newspapers would note who was in town for a wedding, visiting an aunt, or attending a convention. In 1930, the *Bryan Daily Eagle* reported that Corrie Hooks, along with Bessie Reid and two of Corrie's children—Corrie Louisa and Billy—passed through College Station and Bryan. The group was "enroute [*sic*] to the Big Bend country to make a study of bird life preparatory to writing an article for *Nature* magazine."[34]

In a 1932 letter between two of Corrie's children—one in Junction and the other in Beaumont—her daughter wrote, "By the way, I haven't related the fact that Mamma has gallivanted off with Mrs. Reid. They are going to Corpus Christi and points south for the purpose of attending conventions, seeing birds, etc. (mostly etc. probably.) The night before Mamma left, she had a fever of 101, but she went on as she seemed to feel ok the next morning." The letter relayed other family news as well. Clearly, by 1932, Corrie's family knew she could not be restrained from traveling to serve her clubs and explore in nature.[35]

Verna Hooks McLean, circa 1920s. Courtesy Corrie Herring Hooks's family

It is not clear whether Corrie discovered the avian wonderland of High Island before or after her oldest daughter, Verna, married Marrs McLean in 1918. Regardless, once Corrie found her way to the island, she made it her sanctuary and invited other birders to join her there.

Marrs McLean of Beaumont was sometimes referred to as the "second prophet of Spindletop," as he doggedly insisted that the best place to drill for oil was on the flanks of the salt dome, rather than the top. His theory proved to be correct, and he breathed fresh life into the depleted Spindletop oil field in Beaumont. He quickly bought lands and leases around the flanks of salt domes, including the one at High Island. Marrs's investments in High Island and along the Bolivar Peninsula can be traced to as early as 1915 when the first hotel and boarding house he owned was destroyed in a hurricane.[36]

In 1927, Verna and Marrs acquired the famous three-story Sea View Hotel, located on the edge of an oak grove on High Island. The property commanded a view of the expansive marshes and Gulf of Mexico. The hotel had been built by Captain James Taylor Cade in 1897 and was a place of retreat for city-dwelling families wanting to enjoy the beach. According to Melanie Wiggins's book *They Made Their Own Law: Stories of Bolivar*

Peninsula, when oil was discovered at High Island and along the Bolivar Peninsula, the hotel became a "favorite haunt for wildcatters, dealmakers, roughnecks, and other characters."[37]

The McLeans gave the hotel to Corrie and her son Joe Hooks, who ran it until his death in 1941. The hotel was used frequently by family members, friends, and birders and housed Marrs McLean's business associates. In 1947, the grand hotel collapsed in a blaze of glory as it burned to the ground. The small cabins on the property survived the fire, and Corrie continued to use them for many years. In a letter dated May 4, 1960, and postmarked from High Island, she wrote that the "trees were filled with migrating warblers unlike anything I have seen in all my years of bird watching. They poured in by the hundreds with as many grosbeaks, thrushes, and tanagers."[38]

In 2001, Corrie Louise Hooks Pfeil honored her mother and family by donating the three-acre lot that once housed the Sea View Hotel to the Texas Ornithological Society to be used as a bird sanctuary. At the dedication, Corrie's grandchildren who attended recalled fond memories of exploring the natural wonders of the island with their grandmother.[39]

Sea View Hotel, High Island, from family photos. Courtesy Corrie Herring Hooks's family

Through her nature journals and reports, Corrie Herring Hooks helped raise awareness of the value of High Island as an important point in the migration of millions of songbirds every year and as a nesting location for colonial wading birds. Over 350 species of birds have now been recorded on the island, which is just thirty-eight feet above sea level yet is the highest point along the coast between Mobile, Alabama, and the Yucatan Peninsula.

Corrie was a founding member of the Big Thicket Association and is noted for leading frequent trips into the thicket with elected officials and reporters to share the wonders of the woods. She had a deep knowledge of the plants and wildlife of the area and often used her photographs to give presentations to universities, sporting clubs, local civic clubs, and elected officials. Her photographs were a source of inspiration that made the vast, dense forest seem tangible and relatable. Being from an expansive East Texas family gave her credibility when speaking to locals and asking them to join the cause, which was often contentious.

One of the events that helped propel the conservation of these thousands of acres of woodlands came when Corrie's daughter Olga Hooks Lloyd, an avid birder, identified an Ivory-billed Woodpecker on a tract of land along the Neches River in 1966. The Ivory-billed Woodpecker was feared to have been extinct, and ornithologist John V. Dennis was dispatched to confirm the sighting. He spotted a female woodpecker not far from where Olga had reported seeing the species. The Endangered Species Act had not yet been passed; however, priority was being given to conserve lands at the state and national levels, with the highest priority extended to those entities that offered habitat for threatened and endangered species. The sighting of the Ivory-billed Woodpecker reemphasized the fact that the Big Thicket Forest was filled with numerous species of plants and animals that were unique and in peril.

Like her friends Bessie Reid and Connie Hagar, Corrie submitted detailed bird reports from High Island, from her beloved home in Beaumont, and from the Big Thicket. Her observations and accuracy earned her a respected position in the ornithological world. She shared many of her observations with Harry Church Oberholser and pressed him to finish his twelve-thousand-page manuscript on the birds of Texas. Oberholser died before he could his life's work. Corrie doubled her efforts to get

PAGE FOURTEEN

FRIDAY, FEBRUARY 16, 1962

PAINTINGS TO ILLUSTRATE BOOK—Ornithological paintings by the late famed artist, Louis Agassiz Fuertes are being shown by Dr. Frank H. Wardlaw, University of Texas Press director, to Mrs. J. L. Hooks, LEFT, and her daughter, Mrs. Marrs McLean, donor of the fund established to publish "Bird Life of Texas," by Dr. Harry C. Oberholser. The paintings will be reproduced to illustrate the book.

'BIRD LIFE OF TEXAS'

Corrie Hooks; Frank H. Wardlaw, director of the University of Texas Press; and Verna McLean review the artwork for *Bird Life of Texas*, February 16, 1962, *Austin American-Statesman*. Courtesy Katherine Korbell Brown

the book printed but passed away before it was published. Verna and Marrs McLean created an endowment at the University of Texas to pay for the completion and publication of the book in 1974. A student of Oberholser and the nephew of J. Frank and Bertha Dobie, Edgar B. Kincaid Jr., wrangled and edited the book to 1,108 pages of content, photographs, and bird paintings. The attractive two-volume set of *The Bird Life of Texas* is dedicated to Corrie. The Corrie Herring Hooks Endowment at the University of Texas focused for many decades on publishing books related to nature and outdoor exploration. To date, there are more than forty books in the series.[40]

The Corrie Herring Hooks Bird Sanctuary is also part of this dynamic woman's legacy. Most importantly, that legacy has materialized through instilling a love of nature in her children, which seems to have passed to her grandchildren and great-grandchildren. Her long line of family members continues to manage their ranches and lands as mindful stewards. In a correspondence shared by Corrie's youngest granddaughter, she reflects on her grandmother's devotion to nature. "A number of years ago friends of ours gave me a book by Edward O. Wilson titled, *The Diversity of Life*. In it I found a statement he made that to me embodies our grandmother's love of nature and her commitment to its study and preservation: 'And finally I will argue that every scrap of biological diversity is priceless, to be learned and cherished, and never surrendered without a struggle.'" Her granddaughter stated, "I committed that verse to memory and think about it every day. For the eight years, that I was blessed with having her in my life, I saw her live this creed."[41]

THE TEXAS BIRD LADY

Martha Conger (Connie) Neblett Hagar (1886–1973)

Often referred to as the "Texas Bird Lady," Connie Neblett was born in Corsicana, Texas, on June 14, 1886, to Robert Scott and Mattie Yeager Neblett. Connie's father was a lawyer in Corsicana and later served in the Thirtieth Legislature for Navarro County in 1907. Both parents were active in service and clubs that aligned with their deep devotion to the community.

As a child, Connie loved—but did not have a deep interest in learning about—nature. Her first love was music, and she studied voice and piano at Forest Park College in Saint Louis, Missouri, and took postgraduate music training at the University of Chicago and the American Conservatory. Although offered employment as a professional singer, she declined,

Connie Hagar with her dog Patches, 1956. Photo by Edwin Way Teale. Courtesy The History Center of Aransas County

believing such a pursuit to be improper for a Southern lady. More importantly, her mother thought so.

Connie married her neighbor and childhood friend Lynn Brooks. Lynn had a restless soul, which found harbor in the navy a few years before World War I. The young couple moved to New Orleans, Louisiana. Connie basically lived alone, seeing her husband only when he was on leave. As the US entered World War I, Connie returned to her family home in Corsicana to assist with her ill and aging grandmother. While in Corsicana, Connie sang with a quartet at graveside services during and after the war. In the cemeteries, she began to notice nature: first wildflowers and later birds. By the end of the 1918 Spanish flu epidemic, she was the last surviving member of the quartet. That same year, she lost her father to an illness not related to the Spanish flu.[42]

After trying to make her first marriage work for fourteen years, she finally divorced and moved home with her mother in 1921. Shortly after arriving in Corsicana, she met Jack Hagar, an oil and real estate businessman. He slowly won her heart, and they were married in 1924. Jack adored his petite, doll-like bride and took great pride in creating a supportive world where she could do the things that made her happy. Connie suffered from a myriad of health issues, including near blindness, and Jack attended to every detail to keep his wife healthy.

During this time period, Connie and her sister Bert formed the eighteen-member Nature Club, which was affiliated with the National Audubon Society and her local garden club. After formation, Connie started a nature journal that she wrote in every day until she was finally overcome by blindness near the end of her long life. The club took field trips to observe birds, wildflowers, trees, and other wonders of nature. Eventually, the Nature Club gained permission to turn Oakwood Cemetery in Corsicana into a wildflower sanctuary. Connie always liked to visit cemeteries as she traveled to look for native wildflowers and birds. Additionally, she and Bert alternated serving on the Conservation Committee of the Texas Federation of Women's Clubs and often gave presentations together on various topics.

In 1933, Bert's doctor told her she needed some sea and sunshine to ease her nagging arthritis—an affliction from which Connie also suffered. The two traveled to Rockport to spend a month focusing on their health and finally got a chance to see the avian life that had captured their attention in the few bird guidebooks to which they had access. Connie fell in love with the birds and the region.

After her mother passed away, Connie broached the idea of moving to Rockport, and Jack agreed. In 1935, they purchased a motor court with little cottages for rent where they could live and have an income. While living in Rockport, Connie reported more than five hundred bird species in the Aransas Bay area, nearly three-fourths of all the bird species known to exist in North America. She was the first to report the presence of many species in Texas and added more than twenty new ones to the avifauna list.

Many key ornithologists doubted her reports that appeared in the *Gulf Coast Migrant*, a publication started by Rice University Professor George Williams. She never challenged the doubters but would simply respond to their letters with, "Come and see for yourself. I'll be happy to show you around." Come they did. At first to grill her with questions and suspicions, and then later to seek her expertise. Famed ornithologist Harry Church Oberholser made several trips to visit Connie and Jack after he determined she was a reliable resource for information. She sent him her monthly bird reports, which he used in his two-volume *The Bird Life of Texas*.

Roy Bedichek, the well-known naturalist from Austin, heard reports from his friends the Dobies and others that Connie knew everything there was to know about birds along the coast. He thought that was a bold claim, so he paid her a visit. In one afternoon, he was won over after she showed him a nested pair of Mexican Grebes. He had brazenly wanted to call her out as a fraud, but after their time together, he realized she was legitimate. Connie relayed their conversation to Karen Harden McCracken:

> "I have heard people in Austin talk about you, and the people in the Valley, and all the wonderful discoveries you make, and I have said, 'That woman is a faker. Nobody is as good as people say she is.'" That woman [Connie] was laughing, but [Bedichek] went on confessing, "To tell you the truth, I came down here to prove it. Now I am ashamed."
>
> [Connie replied,] "Dr. Bedichek, you are not the first to doubt me. This has been going on ten years. I am used to having to pro-

duce proof for everything I report. You have been far kinder than some who doubted me."

At that, the big man bristled indignantly. "Well, you just send them to me, now. Anytime anyone questions your reports. I can tell them, and I will!"[43]

Indeed, Bedichek became one of her biggest fans. He and his wife, Lillian Bedichek, often visited and invited Connie and Jack for adventures in the Hill Country, including coming to see the Golden-cheeked Warblers. Lillian was a member of the Violet Crown Garden Club in Austin and often reached out to ask Connie to speak to their club about birds, marine life, or any other topic Connie wished to explore.

In 1938, banker and naturalist Guy Emerson, who served on the board of directors of the National Audubon Society, had to see for himself what was really going on in Rockport. After a few days with Connie, he was awestruck.

"As the train pulled away, he settled in his seat and mentally castigated himself. The truth was, he had gone to Rockport reluctantly, not expecting much of Mrs. Hagar. He thought few women were really good birdwatchers, but Oberholser had so insisted and Guy had promised. He had made this trip simply to discharge that promise, and now he was ashamed."[44]

Guy and Connie became lifelong friends and corresponded frequently. From that visit on, Guy sang Connie's praises to anyone who would listen and sent them to see her and stay at the cottages. During her time in Rockport, Connie also formed a lifelong friendship with famed artist and author of multiple bird guidebooks Roger Tory Peterson. As the ornithologist came, so too did a line of other bird watchers. In many ways, Connie became one of the first people to actively promote nature tourism in Texas.

Connie always encouraged her Texas Federation of Women's Clubs friends to come study the birds or flowers along the coast and frequently traveled with them to the Rio Grande Valley. She gladly welcomed visits from Ellen Schulz Quillin and her husband, Roy, as they traveled to study rare plants and birds in the prairies and wetlands of South Texas. Connie's garden club in Corsicana used Ellen's 1928 book, *Texas Wild Flowers:*

Connie Hagar in her heels and dress collecting a bird to band it, date unknown. Courtesy Mary and Jeff Bell Library, Texas A&M University, Corpus Christi, Texas

A Popular Account of the Common Wild Flowers of Texas, as the club explored Northeast Texas.

Dressed in her pressed shirt, flowing skirt, and pearls and with perfectly coiffed hair, Connie would take ornithologists, friends, students, schoolchildren, and anyone else to see the birds. When famed photographer Alfred Eisenstaedt from *Life* magazine arrived to photograph her for a 1956 article, he complained that she did not look or dress like a bird watcher. She invited him to look in her closet to see if she had something more suitable. He found the small closet filled with more skirts and blouses instead of the outdoor gear he thought he might find.[45] The only two accessories that Connie always had with her while birding were her binoculars and her little dogs. Her dogs were her constant companions and kept her safe from snakes as she cautiously walked into the high grasses surrounding the wetlands or in the prairies.[46]

Connie believed strongly in the power of education of children and adults. She thought if people could see, experience, and learn about nature, they would, in turn, strive to protect and care for it. She developed captivating presentations that included taxidermy birds, feathers, nests, eggs, and other objects of nature that allowed her audiences to see and feel the wild things about which she was speaking. Laced throughout her talks was a conservation ethic and call to action to set aside habitat.[47] She gave presentations forty-two times in 1950 and held about the same schedule in the years before and after.[48]

In 1960, the students of Del Mar College in Corpus Christi requested that a day be named for Connie. February 24, 1960, was named Connie Hagar Recognition Day.[49] In the fall of 1962, Jack passed away. Connie was devastated. That same year, the National Audubon Society convened in Corpus Christi largely to be near the coast and to allow the seventy-six-year-old "bird lady" to attend. Audubon surprised Connie with an award that year. According to Karen McCracken, it took a team of people to convince Connie, whose health was failing, to attend and safely deliver her to the banquet. Despite the event being attended by eight hundred people, including US Senator Ralph Yarborough and teams of reporters, the secret of the award was kept, and Connie was humbly surprised.[50]

After Jack passed away, Connie sold the aging motor court and moved into a small house, enlisting a willing team of friends to check on her. The property was later turned into the Connie and Jack Hagar Cottages Sanctuary, which now contains a pavilion and informational panels about the Hagars, along with a bird observation tower and a nature trail.

Connie Hagar died on November 24, 1973, in Corpus Christi after two years of hospitalizations and blindness. She and Jack are buried in a grove of oaks at the Rockport Cemetery overlooking Tule Marsh and Little Bay, where she spent so much time observing the birds and gaining a deep understanding of their daily and annual habits.[51] Connie did more for bird-watching and nature tourism than any other conservationist in

Texas. She figured out the timing of the spring and fall migrations and proved the birds crossed the Gulf of Mexico while other birds followed the land from South or Central America to North America.

Karen "Kay" Harden McCracken (1905–1992) does an excellent job capturing Connie's story in her book *Connie Hagar: The Life History of a Texas Birdwatcher*. Karen was a journalist who wrote, among other things, the "Bird Notes" column for the *Corpus Christi Times* for thirty years. She was also an avid birder and naturalist who interviewed Connie over the course of a year to ensure her story was captured, published, and remembered. Like Connie, Karen wanted to share her observations of nature to inspire others to do the same.

NATURE TOURISM TRAILBLAZER

Madge Lindsay

In a February 2020 interview, Madge Lindsay recounted one of her first and strongest memories growing up in Alabama. "I must have been three years old. My mom, aunt, and I were at my grandmother's house. We walked down a dirt road to some land along the Tennessee River that was getting ready to be flooded for the TVA's [Tennessee Valley Authority] Joe Wheeler reservoir, and hydroelectric power. It was a beautiful pine grove. I remember the smell of pine, the wind in the pine needles, and the sight of those big, beautiful, old trees." That strong connection with the land lasted a lifetime as she spent the majority of her career focused on creating opportunities for people to connect with nature.[52]

A degree in the field of conservation was not on Madge's radar when she attended the University of Alabama with the intent of earning a degree in business administration and management. After her second year of school, she went to work as a secretary for Boeing in Huntsville, Alabama, which was building the first stage of NASA's Saturn moon rocket. Around the same time, she married and started a family. The young couple and their son, Brian, later moved to Dallas, where she volunteered with The Nature Conservancy on a prairie restoration project. After she and her husband divorced, and her son graduated from high school, Madge decided to return to college. In 1991, she earned

Madge Lindsay, 2001. Photo by Larry Ditto. Courtesy Larry Ditto

her bachelor's degree in parks, recreation and leisure studies from Texas A&M University.

"I took lots of botany, natural history, and geology classes. It was so interesting, and fascinating, I was in love with it all. It was like light bulbs going off. You know, when you're older and in school, it's just better, and I could hardly wait to work with what I had learned," she recalled. Madge also mentioned that she was nervous about what would happen after graduation. The worry was for naught. After completing an internship at the Texas Parks and Wildlife Department's nongame program, she was hired by John Herron in the new Wildlife Diversity Program.[53]

In the early 1990s, the Texas Parks and Wildlife Department began tracking outdoor recreation and tourism trends. The data demonstrated that the economic powerhouses of hunting and fishing were declining while activities such as camping, bird-watching, and nature exploration were on the rise. By 1990, the majority of Texans were living in urban areas with less connection to private lands where a hunting ethic could be fostered. The interest in fishing remained stable; however, fewer spaces were available for urban-based citizens to participate in the sport. While sales for hunting and fishing licenses dropped 10 percent from 1983 to 1993, visitation to Texas state parks increased by 40 percent.[54]

This trend was not isolated to Texas but was consistent across North America. Urban centers and suburban communities did not offer the same connection with nature that previous generations had enjoyed. That is not to suggest there was a lack of interest in the outdoors. What people lacked was access and the confidence to pitch a tent, bait a fishing hook, hike a trail without getting lost, or know what to do during an encounter with wildlife. They were certainly less likely to take up hunting without having a connection to the sport in an earlier stage of their life. Young parents were particularly less likely to try a new outdoor skill that required either specialized equipment or the knowledge to explain that skill to their children.

In 1992, Clemson University conducted a survey to capture the attitudes of Americans toward outdoor recreation and tourism in general. The survey found that people who traveled specifically to explore nature or historic sites wanted targeted, meaningful activities through which they could learn about—and be enveloped in—nature while also being physically active. The study also found that nature tourists would go out of their way and spend more money to see an important or interesting bird, flower, or natural phenomenon.[55]

A separate 1994 report from the Eco-Society included data that stated that wildlife viewing was the most popular nature sport in the country, with 76.5 million people participating in the activity. By comparison, hunting had only 14.1 million participants. The US Fish and Wildlife's National Survey of Fishing, Hunting, and Wildlife Associated Recreation pointed to the same conclusion: It was time to embrace nature tourism as an economic driver and popular pursuit.[56]

At approximately the same time, the American Birding Association surveyed its members and found that Texas ranked number one as the place most visited to see birds. Both the American Birding Association and National Audubon Society reported that birders were affluent, savvy travelers who stayed longer in an area if they could find a wide variety of birds and access to multiple birding locations.

According to Texas Department of Commerce data, tourism was the third-largest industry in Texas and generated $23 billion in business in 1993. Governor Ann Richards requested that a state task force be formed to consider what the state could do to enhance nature tourism. The task force was cochaired by Andrew Sansom, executive director of the Texas Parks and Wildlife Department, and Deborah Kastrin, executive director of the Texas Department of Commerce. Madge Lindsay served as the project coordinator, as she and Andrew Sansom shepherded the twenty-member task force through the visioning process to create a plan that could be adapted to urban or rural communities wanting to grow their economies through nature tourism.

Armed with the data and surrounded by innovative minds, Sansom saw birding as the nexus for meeting the shifting public demand, all while intentionally influencing economic growth and increasing access to nature. Madge and consultant Ted Eubanks put forward a unique idea to the task force: launching a birding trail. The concept was to develop and publish a series of maps that included the best birding sites along the coast. Signs identifying the sites would be erected at each of the locations. And so, the Great Texas Coastal Birding Trail became the first trail of its kind in America.

Ted Eubanks and Madge Lindsay with Great Texas Coastal Birding Trail maps, circa 1995. Courtesy Madge Lindsay

Madge credits Ted Eubanks for being both a valuable partner and instrumental in identifying the various locations. Their efforts working on the innovative project paid off, resulting in a $1.5 million grant from the Texas Department of Transportation through a federal program for highway enhancements. The birding trail was later presented the Tourism for Tomorrow Award from British Airways.

In 1995, the first sign along the five-hundred-mile Great Texas Coastal Birding Trail was unveiled at the Connie Hagar Sanctuary in Rockport. Madge coordinated the event and brought in author and artist Roger Tory Peterson, who had been friends with Connie and Jack Hagar.[57] In the 2020 interview, Madge shared that the dedication ceremony was a big hit except for the fact that it was blazing hot that day. "A small speakers' tent that was ordered was not delivered, so Peterson, who was not acclimated to the heat, had to stand in the full sun in his wool sports jacket during his speech. He passed out and had to be rushed to the hospital. A memorable event indeed."[58] In 2020, the program, later renamed the Great Texas Wildlife Trails, celebrated its twenty-fifth anniversary, and it now encompasses nine distinct ecoregions across the state.

While developing the birding trails, Madge partnered with bird watcher Rob Fergus to pull together a "think tank" of birders (included prominent experts Victor Emanuel and Greg Lasley) to work out a plan for encouraging the start of a birding competition loosely modeled after a golfing classic. Coordinating the committee, Madge guided the Texas Parks and Wildlife Department to launch the "biggest, longest and wildest" birding competition in the US. The event would include a giant chrome trophy inscribed with the winning team's name each year. Its entrance fees and donations would collectively pay for bird-specific conservation projects. The Great Texas Birding Classic was hatched.

According to Shelly Plante, nature tourism coordinator at the Texas Parks and Wildlife Department, the birding classic, in its first twenty-four years, funded more than $1 million in projects that support nature tourism, avian habitat restoration, and land conservation. Shelly has been the keeper of the classic and the Great Texas Wildlife Trails since Madge hired her as an intern in 2001.[59] For the first ten years, the Texas Parks and Wildlife Department partnered with the Gulf Coast Bird Observatory under the direction of Cecilia Riley to manage the program. Once the birding classic grew to include more than just the coast, the partnership evolved and now resides in the Outreach and Education Department at the Texas Parks and Wildlife Department under the direction of Shelly Plante.[60]

With the success of the Great Texas Coastal Birding Trail and the Great Texas Birding Classic, Texas Parks and Wildlife was ready to try an even bolder approach to enhancing nature tourism. In 1996, the concept of the World Birding Center in the Rio Grande Valley was devised. Madge championed the idea to Andrew Sansom, and he assigned her to organize the effort. As the project gained momentum, Madge moved in 1998 from Austin to Weslaco in the Rio Grande Valley to coordinate the local task force.

The task force, in partnership with the Texas Parks and Wildlife Department, set the parameters of what each birding center site should offer in order to be awarded funds and carry the brand of "World Birding Center." When the task force shared the nature tourism economic impact data with the communities in the valley, it didn't take long for the cities to see the value.

In a September 1997 McAllen, Texas, *Monitor* article, Gillian Swanson reported, "The battle for the birds has begun. Weslaco this week publicly

Shelly Plante and Martha McLeod at the Great Texas Birding Classic awards in San Antonio, 2017. Courtesy Jennifer L. Bristol

announced its hopes of landing the multimillion-dollar World Birding Center. Harlingen and McAllen already have revealed their wishes to win the facility, which will house exhibits, maps, guides, and up-to-date bird sighting information for expert birders, tourists and community members alike."[61]

Indeed, the competition became heated. Cities pulled together packages that included land, matching funds, private partnerships, and community input. With $15 million on the line, there was a lot of jockeying for position, with some cities tossing their names into the hat just hours before the deadline. The project grew in scope to include a central education center and six satellite centers. The most popular idea was to have a centrally located headquarters, which put Weslaco at the top of the list, particularly since land was already being purchased there for the purpose of creating a new state park.[62]

Then the Texas legislature got involved. In an April 1999 *Monitor* article titled "World Birding Center Will Nest in Mission. Surprise Decision Comes After Legislature Intervenes," Cheryl Smith reported,

> The Texas Legislature's appropriations committee took the bird by the beak Thursday when members voted to award the coveted World Birding Center's headquarters to Mission, stripping Texas Parks and Wildlife Department of any decision-making authority on the site. In a surprise move, the House of Representatives Appropriations Committee's Conference Committee agreed to provide the Texas Parks and Wildlife Department with $35 million in additional funds on the condition that Mission be selected as the headquarters for the World Birding Center. "About $5 million of the approved Parks and Wildlife funds are earmarked for the center," said State Representative Ismael "Kino" Flores, D-Palmview, who serves on the House Appropriations Committee. Flores pushed for Mission to be awarded the WBC headquarters.[63]

The article goes on to mention that the mayor of Mission was Representative Flores's second cousin.

The decision by the legislature was a surprise and a blow to the Texas Parks and Wildlife Department and the task force. It undid much of the goodwill and trust that had taken years for Madge to build. Instead of crumbling, Madge and the task force pressed on. The Texas Parks and Wildlife Department moved forward to transform two of its parks into World Birding Centers and add another park under that designation.

The selected headquarters was to be at Bentsen–Rio Grande Valley State Park in Mission instead of the new park in Weslaco. The original 587 acres were donated to the state in 1944 by the Bentsen family for the purpose of a park. The ebony-forested park opened to the public in 1962. Additional purchased and donated lands were added for the new World Birding Center headquarters complex. Today, much of the park resides behind the border wall but is still accessible to visitors willing to walk, bike, or take the tram to enjoy the bird blinds and ADA accessible hawk-watch tower.[64]

As the City of Weslaco and the Texas Parks and Wildlife Department continued their plans to add the new bird-focused park, they requested

Celeste Garcia at Estero Llano Grande State Park and World Birding Center, 2022. Courtesy Martha Garcia Martinez

due to lack of rain. When she was given the green light to pump water from the nearby channel to fill the wetlands, they instantly came to life with birds, frogs, and alligators. As a biologist, she reported how rewarding it was to see the first frog in a space that had been a dry plowed field for so many years.[68]

In addition to building partnerships and overseeing the construction of the facilities, Martha engaged volunteers to help develop the site-specific curriculum. Once the park was open, she encouraged the schools to utilize it as a living classroom where kids could see, touch, and feel nature as they learned about science, history, and other topics. Martha advocated to upgrade several of the RV sites so year-round park hosts could live and volunteer at the park, thus increasing the number of skilled helpers to lead nature walks and deliver programs. The initiatives and programs Martha established carried on even after she relocated in 2014 to become superintendent at Stephen F. Austin State Park in San Felipe.

Martha was the first Hispanic woman superintendent within the state park system. She celebrated her twenty years of service in 2022 and shared that her daughter, Celeste Garcia, was an intern at Estero Llano Grande State Park and World Birding Center that same year.

The last of the centers opened in 2012 with the addition of the Hugh Ramsey World Birding Center Site in Harlingen. The nine locations span the Rio Grande Valley, from Roma to South Padre Island, and continue to be a mecca for birders from around the world. By 2010, an estimated $125 million had been infused into the valley's economy by nature tourists, with a primary focus on bird-watching.

According to the Texas Parks and Wildlife Department, visitation at the three state parks that are identified as World Birding Center sites increased by 68 percent—from 660,000 visitors to 1,106,000—between 2019 and 2021. This rise during the COVID-19 pandemic demonstrates the public's need and desire for more outdoor opportunities for local residents and travelers alike. Even more striking, the out-of-town visitors came primarily from the United States rather than other countries since travel was restricted for much of that time. With the Rio Grande Valley having the eighth-largest population in the United States, and having lost 95 percent of the native habitat, each of the World Birding Center sites is an important piece of the conservation matrix.

After four years of wrangling the multimillion-dollar, multilocation project, Madge Lindsay was ready for a change. In 2002, she left Texas to become the state director of Audubon Mississippi in Holly Springs. In that role, she helped launch a conservation plan that included birding trails along the gulf coast and Mississippi River. While there, she expanded the Strawberry Plains Audubon Center and helped launch the Coldwater River Watershed Initiative and Hummingbird Festival. *Birders World* magazine named Madge one of its seven "Heroes of the Birds" in 2002.[69]

In 2005, Madge partnered with Cliff Shackelford, Mark Klym, Clemente Guzman, and Sid and Shirley Rucker to publish *Hummingbirds of Texas* through Texas A&M University Press. Madge retired to the Davis Mountains, where she, Cecelia Riley, and Linda Hedges volunteered to continue the Davis Mountains Hummingbird Celebration which had started in the 1990s. Each year, the festival brings in several hundred people to enjoy the jewels of the sky and learn about the ecological wonderland of the Davis Mountains region.[70]

When I asked Madge what advice she would give to young women wanting a career in conservation, she replied, "Once you discover nature, you really can't let it go. It's just something you don't want to lose. Working in conservation is probably the most rewarding career you can have." Madge certainly found a way to create a career in an agency that had few women when she began there. Nor did she shy away from the grand egos involved with establishing the World Birding Centers in the Rio Grande

Madge Lindsay with Shelly Plante at the Rio Grande Valley Birding Festival, 2019. Courtesy Jennifer Bristol

Valley. "Having a healthy sense of humor certainly made the hard days easier."[71]

It is hard to calculate the legacy of Madge's contributions to conservation, as millions of people enjoy the World Birding Centers, Wildlife Viewing Trails, Greater Texas Birding Classic, and Hummingbird Celebration each year. Her legacy reaches beyond the borders of Texas, as the birding trails inspired more than twenty other states to implement their own versions. The World Birding Centers inspired fifteen additional nature centers throughout the Rio Grande Valley. The focus on nature tourism and birding has also motivated many state and local parks to add bird blinds, hawk-watch towers, and boardwalks for viewing the birds. Thanks to Madge's innovation and ability to build coalitions to support nature tourism, billions of dollars continue to be infused into the Texas economy every year.

Madge observed, "All of this happened with the help of enthusiastic people I have had the pleasure of working with, many very smart women and men coworkers, who love nature as much as I do."[72]

THE WOMEN OF HOUSTON AUDUBON

Winnie Burkett, Joy Haynes Hester, and Helen Drummond

In 1900 the population of Houston was around 44,000 people. By 1920, during the height of the first oil boom, the population swelled to around 138,000. The influx of people, businesses, houses, oil refineries, chemical plants, shipping and transportation hubs, and roads was beginning to have a noticeable impact on the avian, aquatic, and terrestrial communities. Habitat loss was massive as the prairies were turned into agricultural fields, forests were logged for lumber, and wetlands were drained in the name of progress.

Concerned about what he was seeing, Joseph Heiser started the Houston Outdoor Nature Club in 1923 to study the flora and fauna, while also raising awareness through outreach and education. He engaged the Houston and River Oaks Garden Clubs in a variety of conservation projects, including working with Catherine Emmott to establish Memorial Park.

In the late 1920s and early 1930s Joseph and the Outdoor Club worked with Bessie Reid to secure the nesting habitats for the Roseate Spoonbills on the Vingt-et-un Islands and hire a game warden to protect the rookery. In 1948, he also lent his expertise and enthusiasm to the Texas Garden Clubs' Audubon Nature Camp in Kerrville as one of the instructors. Joseph's interest in birds parlayed into inspiring the formation of the Houston Audubon Society.

By 1960, with the population of Harris County at 1.2 million people, the bayous were being channelized, the bays were being dredged to accommodate larger ships, NASA was blasting off, and industry along the waterways was staggering. In 1969, with support from the Houston Outdoor Club, eighteen citizens gathered to form the Houston Audubon Society. Among the first board of directors were Carl H. Aiken, Dirk Hagemeyer, C. E. Hall, Terry Hershey, Norma Oates, Harvey Patton (chair), T. E. Pulley, W. G. Scheibe, Mary Sears, Margaret P. Sullivan, and Armand Yramategui.[73]

The organization quickly rallied to preserve a piece of the Cypress Swamp and Trinity River, and thus began their work in advocacy for habitat and birds. In 1975, conservationist Edith Moore bequeathed to Houston Audubon 17.5 acres of land that included the log cabin that she and her husband, Jesse, had constructed, thus establishing the Edith L. Moore Nature Sanctuary.

In the 1980s Houston Audubon began acquiring properties on High Island and Bolivar Peninsula. Winnie Burkett, biologist and sanctuary manager, told David Todd in a 1999 interview that buying the first four acres of land for the Smith Oaks Sanctuary was remarkable because most Audubon societies around the country were not taking on sanctuaries at the time. She pointed out that even within the Houston Audubon board, there was conflict between those who wanted to put funds toward it and those who thought they should only take on lands that were left to them. However, the critical habitat of the live oak forests on the highest point along the Texas Gulf Coast of Mexico had to be saved from development, so the board took out a loan to buy the lands.

Winnie recollected, "Houston Audubon board passed resolutions that they couldn't use any of the organization's money to buy this land. They had to raise separate funds. And so, they had, you know, cake sales and they had various and sundry things. But nothing raised a whole lot of money . . . so they started charging admission. And that was a very novel idea at the time. In fact, a lot of people objected to that. But, that's the way that they paid the mortgage off."[74]

As the 1990s marched along, Houston Audubon continued to take on properties left to them or purchased critical habitats outright. There was not always rhyme or reason to the acquisitions, but it seemed to work. In the meantime, thanks to Winnie, Houston Audubon established the Hawk Watch at Smith Point, and partnered with seven other entities to form the Gulf Coast Bird Observatory, and continued their volunteer-led education programs.

Always looking for worthy volunteers, longtime Audubon member Peggy Boston met Joy Hester while birding at the Hana and Arthur Ginzbarg Nature Discovery Center. The two exchanged numbers and, as Joy later recalled, "that is when I started down the slippery slope of volunteerism and quickly and happily found myself in up to my neck!"[75] Peggy

Winnie Burkett and the Muddy Buddies Houston Audubon Birdathon team at High Island, date unknown. Courtesy Helen Drummond

mentioned to Winnie there was a new volunteer who was also a well-respected real estate and contract attorney. Within weeks Winnie Burkett was calling Joy with land acquisition questions.

Joy Haynes Hester grew up along the border of Texas and Mexico in Eagle Pass and recalls that her childhood was not filled with fond memories of nature. "Growing up in West Texas, in farming and ranching country, nature was the enemy in many ways. My grandfather was a Pan-

Catherine Hannah (*left*) and Joy Hester volunteering at a Houston Audubon Sanctuary. Courtesy Houston Audubon

handle rancher, and my dad grew up on that dry and hostile land. Families there survived the Dust Bowl and major droughts. DDT and other chemicals were welcomed. I didn't find my love of nature really, until I had children of my own." She laughed as she continued, "Like many small-town girls, I married right after graduating from high school and had two sons soon thereafter. We moved to Louisiana and when they started school, I started college. Actually, my husband returned to college then too. At night we would all be studying and reading. It was a great time. I think it really helped my boys excel in school."[76]

Later the family moved to Austin and Joy took a job at Austin Independent School District. "I went as far as I could in that career but since I didn't have a PhD, I couldn't go any further. I knew I didn't want to get a PhD. My younger brother was a lawyer and I thought, 'If he can be a lawyer, why can't I?' So I decided as my boys were heading off to college that I would go to law school. My son and I used to bump into each other between classes at the University of Texas. That was in 1979."

After graduation, Joy proposed that they move to Houston since the boys were grown and out of the house. Her husband landed a job at NASA, and she went to work for Texas Eastern Transmission in downtown Houston. "I dealt a lot with contracts, which I thought was boring at the time but learned to enjoy and that led to real estate law which I loved. I eventually ended up as a partner in a small law firm, representing corporations and school districts in real estate and contract matters. It was all good until our younger son came home from Yale with a mysterious illness that came to be known as AIDS. Our other son and his wife, both lawyers, moved to Houston from Washington, DC, and helped us through it, emotionally and financially. It was then that my values and my career path changed. My husband went back to work at NASA, but I lost all interest in the practice of law and withdrew from the law partnership. And then Nature rescued me." Sadly, their son passed away.

In 1999, after volunteering and assisting with a land transfer to establish the Carolyn Raizes Davis Sanctuary on Chocolate Bayou, Joy was asked to join the Houston Audubon board by Jeff Mundy, a well-known trial lawyer who focuses on wildlife protection, water rights, and environmental justice. Joy said yes and "slid a little further down the slippery slope." The organization was in a transition period as it cycled through several executive directors in a short time. Joy stepped forward to offer her services as interim executive director.

As executive director, she dove headfirst into one of the most complicated land deals Houston Audubon, or any other conservation organization, had ever entered into. Most people would have run for the hills when approached by a bankruptcy trustee with the idea to buy 650 acres of marshland that was divided and platted into 4,126 separate lots. However, the mix of coastal prairie and marshes that rimmed the bay side of Bolivar Peninsula was too magnificent a bird habitat to pass up.

Reflecting on the project, Joy shared, "It wasn't just that there were 4,000 plus lots. Some of the lot lines extended well into the bay. In Texas, if the coast has reclaimed the land, it becomes the property of the state. Sid Bouse from Coastal Surveying waded deep into the marshes and bay to complete each survey. Winnie Burkett helped identify which lots they should focus on next, and they slowly ticked off each one over the course of three years. Most of the lands were purchased with money from a US Fish and Wildlife grant from the North American Wetlands Conservation Council. The rest of the money came from over 1,000 donations to Houston Audubon from more than thirty states."

Joy continued to reflect, "It was complicated, but I just happened to be in the right place at the right time. The staff was so supportive and everyone, especially the trustee of the properties, wanted to find the solutions to see the project through. Jeff Mundy called me at the end of every day to check in. Being an executive director of a nonprofit is sort of a lonely space. Jeff calling every day was a gift." When the Horseshoe Marsh project was finalized in 2003, Joy recalled, "There's no better feeling than looking out on a big piece of conserved land, teeming with wildlife, and say, 'I got to be part of saving that.'"

Horseshoe Marsh was not the only land deal Houston Audubon had going on during Joy's time. There were several pieces of land that had been left to Houston Audubon that did not meet their sanctuary standards, and Joy felt strongly that a plan was needed to assess all the properties. Happily, her dream of a land master plan for Houston Audubon was eventually achieved. Joy also saw the need for Houston Audubon to take the lead in collaboration with other conservation organizations in the region. "I felt if we worked collectively, we could achieve more and be more strategic with what lands were set aside or what issues we approached public officials with."[77]

After three years, Joy stepped down as executive director and went back onto the board. They hired Gina Donovan as the next executive director and Joy was soon elected board president. In 2004, the board voted to name a new Bolivar Peninsula land acquisition the Mundy Marsh Bird Sanctuary in honor of former board president Jeff Mundy. Joy reflected that it would be a constant fight to conserve the lands along the Bolivar Peninsula and High Island, as developers are always looking for ways they can squeeze in more houses or build a resort, despite the fact that it is a hurricane-prone barrier island.

Under Joy's and Jeff's leadership, Houston Audubon added more than a thousand acres of land. Joy also mentioned that she was proud to support staff member Flo Hannah in adding the Natives Nursery at the Edith L.

Joy Hester, Flo Hannah, and Jeff Mundy at the Houston Audubon High Island Sanctuary, date unknown. Courtesy Houston Audubon

Moore Nature Sanctuary as a means to generate native plants for community partners to use in restoration projects.[78] Along the way, Joy joined the board of the Texas Land Trust Council and encouraged Houston Audubon to complete the accreditation process, which they achieved in 2017. She was elected to serve on the National Audubon board and served on the Audubon Texas board during that organization's restructuring.

When I asked Joy what advice she would give to young women, she replied, "Don't wait to be asked. If you want to be a leader, step forward, let people know what you want. I see too few women taking leadership roles in the conservation world. In the fifty years of Houston Audubon there have only been four women board presidents and Jeff Mundy recruited three of those four. Don't be afraid. Ask for help along the way. But get involved."[79]

Joy was on Houston Audubon's board of advisers when they hired the first African American woman executive director, Helen Drummond, in 2012. Helen was not just the first African American woman to lead Houston Audubon but the first to lead any conservation organization in Texas and one of the first, if not the first, in the country.

Helen's family moved to Houston when she was a toddler, along with her three older siblings. She recalls that her mother often teased that Helen was probably the healthiest of the kids because she was always outside, playing in the dirt. Whether she was exploring her grandparents' farm in Virginia or Brays Bayou in Houston, Helen loved being outdoors and all things water. "I was in awe and mystified by nature, especially water. I loved to fish, swim, or just splash around in a creek or any body of water. I was a child of the wild."[80]

She attended Hampton University, a research university located in coastal Virginia, to study marine and environmental science. Helen's passion for helping others and preserving nature fueled her initial research interest and earned her a grant from the National Institutes of Health to study waterborne illnesses as an undergraduate at Hampton. Helen earned her master of science in environmental management from the University of Houston–Clear Lake, and in 1993 she went to work for the City of Houston as a wastewater compliance officer.

After three years she changed employment and went to work as a water and sediment team leader for the newly created Galveston Bay Estuary Program, jointly administered by the Environmental Protection Agency and the State of Texas. She was on the cutting edge while the project was moving from planning to implementation. As the program evolved and expanded the public-private partnerships, Helen worked her way up to program director, where she thrived for eleven years. "I really enjoyed building coalitions, engaging the stakeholders, and building consensus to address the greatest challenges facing the region, including habitat loss."[81] Habitat conservation was the highest priority for the organization, but large-scale projects often required federal funds and a nonfederal match in addition to funds for due diligence. In 2010, Helen created the Galveston Bay Conservation Assistance Program to facilitate their partners' ability to identify and close on land conservation projects.

Under her leadership, the Galveston Bay Estuary Program won ten national awards, including several Coastal America awards, the highest environmental award of its kind, given by the president of the United

Terry Hershey (*left*) and Helen Drummond, Houston, 2015. Courtesy Helen Drummond

States. During her tenure, Helen served on the Association of National Estuary Programs' Board of Directors, the Texas Sea Grant's Advisory Committee, and the Houston-Galveston Area Council's Natural Resources Advisory Committee, among other boards and committees.

Helen recalls that limited public access to nature, especially to Galveston Bay, was a significant concern of the local conservation community and an issue that touched her deeply. Houston Audubon was one of the estuary program partners. Helen recollects Judy Boyce, former executive director for the Hershey Foundation and a Houston Audubon board member, asking her whether she would consider serving on the Houston Audubon board. While Helen was not in a position to serve on the board at the time, she was positively affected by Judy reaching out to her. Helen admitted she was not a birder at the time; however, she deeply believed that "birds are ubiquitous connectors to nature and have the ability to create opportunities to engage people from all walks of life."[82] When she learned that Houston Audubon was searching for a new executive director, she saw it as an opportunity to pursue "the people side" of her conservation interests and threw her hat in the ring.

Helen was hired in 2012, and her first order of business was to create a long-range strategic direction for the organization. Houston was noted to be one of the most diverse cities in the United States, and the Census at the time showed that Harris County had surpassed 4 million people.[83] That was 2.3 million more people than when Houston Audubon was founded. As Helen engaged the board and stakeholders in the visioning process, she was able to shift the focus from rapid expansion of additional sanctuaries to the sustainability, restoration, and improvement of those properties.

Part of that vision was to shore up inholdings and acquire lands adjacent to the existing sanctuaries where possible. Under Helen's leadership, Houston Audubon increased its land holdings from thirty-four hundred acres to over forty-one hundred acres. The organization made significant investments to improve public access and enhance the visitor experience within sanctuaries, and it established a land and facilities stewardship endowment for the lands it owned.

The Bird-Friendly Communities program was also a pivotal element of the new strategic direction, and Helen recruited Sarah Flournoy, former director of the Nature Discovery Center, to join the team to create the community-centered urban conservation program. The Bird-Friendly Communities initiative creates important corridors of healthy habitat in the greater Houston area, provides accessible opportunities, and unites the community through a shared appreciation of nature.

In 2013, utilizing the National Park Service Rivers, Trails, and Conservation Assistance Program, Houston Audubon and a coalition of diverse stakeholders developed a plan to expand nature-based recreational opportunities and environmental education and outreach for High Island and the Bolivar Peninsula. The plan was called the Bolivar Peninsular

Nature Trail, and in 2014 Houston Audubon began master planning for its High Island sanctuaries. When restoration funding from the BP Deepwater Horizon oil spill was announced, Helen saw it as an unprecedented funding opportunity for large-scale projects and began developing proposals that included projects from the Bolivar Peninsula Nature Trail and the High Island master plan. Winnie and a few others advocated to keep a canopy walk concept in the master plan, but they never really thought it would be funded.

When they shopped the master plan around to various funders, Kathrine G. McGovern of the John P. McGovern Foundation thought the canopy walk would be a wonderful addition to the sanctuary. The foundation sponsored the design and building of the $1.8 million canopy walkway. The fully accessible metal walk slowly rises from the parking lot to eighteen feet into the coastal oaks, before it splits in two directions and ends overlooking two ponds, providing commanding views of three colonial waterbird rookeries. To accommodate the increase of visitors at Smith Oaks Bird Sanctuary, Houston Audubon also improved the parking area and installed a state-of-the-art, accessible restroom, a delight to all birders.

The Kathrine G. McGovern canopy walkway, 2021. Courtesy Houston Audubon

"The Kathrine G. McGovern canopy walkway is a first for Houston Audubon and for the region. High Island is one of the most popular birding destinations in the US during the spring and thanks to the generous support of the John P. McGovern Foundation, visitors will experience nature and a dazzling parade of birds on the elevated canopy walkway," said Helen. "This unique experience has already begun to attract more birders, photographers and nature enthusiasts to the Upper Texas Coast."[84]

Helen mentioned she was also proud of Houston Audubon's Lights Out for Birds Program, which started in 2017 after a large collision event in Galveston. Thanks to the Cornell Lab of Ornithology and their partners BirdCast bird migration forecast maps, Houston Audubon conservation staff were able to use data driven science to predict migration for the region and issue alerts to the community when chances of high migration coincided with weather events.

A 2019 study found that Houston was the second most dangerous city for migrating birds in terms of light pollution and bird strikes with glass-intensive buildings. In 2020, Houston Audubon partnered with Audubon Texas, National Audubon, the Cornell Lab of Ornithology, Texan by Nature, and organizations across the state to expand the program into Lights Out Texas. Through the Lights Out Texas initiative, Houston Audubon was able to convince the City of Houston to pass a resolution supporting the initiative and encouraging city buildings to participate. Then, in April 2021, "Mayor Sylvester Turner proclaimed April 19–May 7, 2021 'Lights Out Nights' in Houston, encouraging all businesses, residents, and building managers to turn off non-essential exterior lighting from 11 P.M. to 6 A.M. during the period of peak migration for birds through Texas." The mayor went on to say that city buildings and facilities would participate in the program.[85]

Helen stated in our interview that she would like to see more policy at the local and state levels that protects habitats and birds. "Staff, board members, and volunteers all move on eventually. We need to institutionalize and normalize the protection of species and valuing the nature around us." She went on to note that COVID-19 certainly highlighted the need for more access to the outdoors at all levels, from neighborhood pocket parks to National Wildlife Refuges and everything in between.[86]

Helen Drummond birding with her daughter Gabby Drummond, 2022. Courtesy Helen Drummond

As the population of Harris County surpassed 4.8 million in 2020, the pressure to conserve open space for people and wildlife is greater than ever. Helen shared that reaching a more diverse population is also a value of the organization in light of the area's growth. One way they achieve that is via the passion and creativity of Mary Ann Weber, the education director, and an army of staff and volunteers, Houston Audubon delivers programs to approximately thirty thousand kids and adults each year. Helen also shared that in addition to the schools they serve, the most diversity they see is in their programs at Houston Audubon's Raptor and Education Center. The Raptor Center is home to Houston Audubon's eleven raptors and other animals that feature in their education programming.

When I asked Helen what challenges she has faced or what advice she would give to young women, she replied, "Be ready to be second guessed. Especially if you are a woman of color. I can't tell you how many times I've said an idea to a group and not ten minutes later a male says the same thing and it's the best idea anyone has ever heard. But you just got to laugh it off and move on." She added, "Identify allies. Get a good support team and don't be afraid to ask for help when you need it. There are those that will always make you feel like you shouldn't be at the table. Know your value and make sure your purpose and passion line up—then no one can question if you should be at the table."[87]

When asked whether she is seeing an increase in diversity in conservation leaders, she laughed and said, "It's been slow, but it is happening. I encourage where I can, and I am inspired by the young Black leaders moving in the conservation space today." Corina Newsome, one of the founders of Black Birders Week whom Helen befriended, is one such leader on the move. Of course, the person Helen has influenced and encouraged the most is her own daughter, Gabriella. "Gabriella has accompanied me on outdoor ventures from Galveston Bay to birding in Houston Audubon sanctuaries since she was four. We enjoy exploring outdoors as a family."[88]

The tangible part of Winnie, Joy, and Helen's legacy can be found in the forty-one hundred acres of land at seventeen sanctuaries that they helped purchase and maintain. Through careful management of Houston Audubon, Joy and Helen have been able to secure the funding and support for avian and conservation education programs for thousands of children and adults each year. The intangible part of their legacy becomes evident when considering how many of those people have made changes in their life to be more conservation minded and more aware of their natural world.

WOMEN'S INFLUENCE IN FORMING THE STATE PARK SYSTEM IN TEXAS

As mentioned in chapter 1, throughout the 1880s and 1890s, women's clubs were forming across Texas in towns both large and small. During that same time, the city of Galveston was becoming a boom town and the main port for Texas farmers and ranchers to export cotton, cattle, and other goods. It was also a hub of women's clubs, as the wives of wealthy merchants sought to construct a more meaningful reality for themselves beyond being reliable mothers or dutiful daughters. Though they were still not able to participate in the political process, the women's clubs were spaces where they could influence change for social issues such as education, sanitation, prohibition, homelessness, criminal justice reform, and historic preservation.

Two Galveston women, Betty Eve Ballinger (1854–1936) and her cousin Hally Ballinger Bryan Perry (1868–1955), met in the library of the Ballinger family home, known as the Oaks, to discuss their concerns about the diminishing memory and fading fighters of the Texas Revolution. Betty's maternal grandfather, William Houston Jack, had fought in the Battle of San Jacinto and later became a prominent lawyer. Hally had the distinction of being the grandniece of Stephen F. Austin, one of the first Anglo colonizers in Texas.[1]

In the spirit of being dutiful daughters, the two wanted to honor the memory of those who had fought in the revolution, as well as their families. In 1891, they traveled to Houston to meet fourteen other distinguished women in the home of Mary Jane Harris Briscoe. Mary Jane was the wealthy widow of Texas patriot and businessman Andrew J. Briscoe. Also in attendance was Mary Smith Jones (1819–1907), the widow of Anson Jones, who was the last president of the Republic of Texas.[2]

The women collectively agreed to take action and form an association first called the Daughters of Female Descendants of the Heroes of '36, which later became the Daughters of the Republic of Texas. The charter for the club was not ratified until 1895 when the Texas Veterans Association formally requested that the Daughters of the Republic of Texas carry

Veterans of the Texas Revolution, circa 1895. Courtesy Texas Parks and Wildlife Department

forward their memory and honor the past. Mary Smith Jones was the first president of the new organization.[3]

The first actions of the Daughters of the Republic of Texas focused on preserving documents, collecting stories from the aging veterans and their families, and organizing celebrations that honored specific dates relevant to the establishment of Texas. Deeply committed to the power of education, they also influenced the teaching of Texas history in public schools.

Knowing time was of the essence to capture the fading memory of the Battle of San Jacinto, they worked with veterans to map the grounds where the final battle between the mighty Mexican Army and the rebellious Texans took place. After several failed attempts to secure funding, they finally found a willing legislative sponsor in Senator Waller Thomas Burns of Houston to put forward a bill to establish the battlefield as a state-funded memorial park. In 1897, with political support from Senator Burns and the constant persuasion of the Daughters of the Republic of Texas, the Texas legislature appropriated $10,000 for the purchase of the battlefield, thus making San Jacinto the first state-owned memorial park in Texas.[4] By 1907, the park consisted of 336 acres, with lands being added over time to encompass the 1,200 acres of the current historic site.

After the success of the establishment of the San Jacinto Battlefield Park, the Daughters of the Republic of Texas set their sights on additional locations associated with the revolution, including the Alamo, the Fannin and Gonzales battlefields, Refugio public square, and Washington-on-the-Brazos. These additional properties sparked the notion to follow the national park model to create a state park system.

The idea and desire for parks came late to Texas. Following national trends, the first parks consisted of cemeteries such as Glenwood Cemetery in Houston (established in 1871) and Oakland Cemetery in Dallas (established in 1891). The next logical progression for the concept of parks was to establish memorial parks to honor the Texas Revolution,

Congress of Mothers of Texas–San Antonio chapter, 1908. Cofounder Mary Eleanor Brackenridge seated in front, third from right, in dark dress. Courtesy University of Texas–San Antonio Libraries Special Collections

Civil War battlefields, and other culturally significant places that were relevant at the time. While the establishment of those types of memorial parks was underway, Texans also acknowledged their desire to have parks for living people as well as places to honor the departed.

Additional women's clubs were emerging across the state, and in 1897 twenty-one clubs gathered in Waco to form the Texas Federation of Literary Clubs. The name was changed in 1898 to the Texas Federation of Women's Clubs when members agreed to expand their focus beyond education and enrichment to include community service, arts, welfare of the poor, support of Native Americans, and parks and playgrounds.[5] Women of color were initially included—and then later excluded—from the Texas Federation of Women's Clubs. In 1905 they formed their own powerful network known as the Texas Federation of Colored Women's Clubs.[6]

The devotion to education and literacy by the women in both federations cannot be overlooked. More than 85 percent of Texas public libraries are the result of both federations. An equally important component to consider is the aspiration of the women in both federations to have the power of the vote and become a part of the political process. Alongside their sisters in other states, they organized, wrote letters, marched, and held luncheons. Texas granted the right for women to vote in the state primaries in 1918 and, in June 1919, became the ninth state—and the first

Men looking at material posted in the window of the headquarters of the National Association Opposed to Woman Suffrage. Photo by Harris and Ewing, circa 1910–20. Library of Congress Prints and Photographs Division, Washington, DC, http://hdl.loc.gov/loc.pnp/pp.print

Southern state—to ratify the Nineteenth Amendment, which enfranchised women.[7]

National laws, state poll taxes, and local voting intimidation laws and practices obstructed men and women of color from voting for decades. Many white women were forbidden by their parents or husbands to vote. Often they were forbidden to work outside the home, or run for office, which slowed the social transformation the suffragettes had envisioned. There also remained a contingency of women who believed it was not the place for women to enter into decision-making roles outside the confines of raising and educating children or tending to their families.

With the end of World War I and recovery from the Spanish flu epidemic, industrialized countries roared into the 1920s. New innovations infiltrated all facets of life, including education, farming methods, and social constructs. Women's fashions were greatly altered during the 1920s and 1930s, which allowed women to move more freely in the outdoors. The new fashions and the activities they supported mirrored a national obsession with fitness and health. New innovations in the home also freed women from the long, arduous labors of preparing every meal

Forest of oil derricks south of Beaumont, Texas, 1901. Photo by Trost Studio, Port Arthur. Courtesy Library of Congress Prints and Photographs Division, Washington, DC

for their families and cleaning. Leisure time for men and women was no longer exclusively the privilege of the wealthy. Middle-class families wanted to vacation at national parks, lodges, and lake cabins or send their kids to summer camp. State parks were the next logical step for fulfilling the growing demand for outdoor recreation.

Transportation was also changing, and the post–World War I automobile age arrived in Texas and saw fertile ground. Farmers and ranchers embraced trucks to haul their goods to smaller markets, rail yards, or shipping hubs. Merchants and salesmen discovered a faster, more efficient transportation system to sell their wares. The growing middle class recognized the automobile as a means to transport them out of the hot, dusty, and often smoky cities to the countryside to spend their newly found leisure time. All these new automobiles required gasoline and motor oil, a resource that ran deep in Texas.

The new automobile age, however, lacked two things in Texas: good roads and public points of interest beyond cemeteries and historic sites. Inspired by the national park movement, modern travelers wanted places to escape into the wilderness, marvel at nature, canoe down a river, swim or fish in a lake, hunt game, or reunite with their family and friends. The Texas Federation of Women's Clubs advocated for improved roads for recreation and, more importantly, to transport rural schoolchildren safely to community schools and provide them with better access to local medical services. As early as 1915, the Texas Federation of Women's Clubs formed a committee under the Rural Life Division to advocate for improved roads.[8]

One of the many men who recognized the potential and understood the necessity to improve transportation was Governor Pat Morris Neff. Pat's parents, Isabella and Noah Neff, built a home in Coryell County along a clear, cool spring they called Neff Spring. Life was not easy on the Texas frontier, where Comanche were always on the lookout for homes to raid and horses to liberate from the pioneers who penetrated ever deeper into their homelands. Despite the hardships, Isabella loved the freedom of country life, and she deeply loved the land. She and Noah welcomed nine children into the world; Pat, born in 1871, was the youngest of the brood.[9]

Pat Neff earned an undergraduate degree from Baylor University and a law degree from the University of Texas. In 1899, he entered the Texas

House of Representatives and, in 1903, became the youngest Speaker of the House. He took a break from politics in 1905 to practice law and attend to his family in and around Waco. In 1920, he saw an opportunity to run for governor as a strong advocate of Prohibition and reform. In his campaign, he covered more than six thousand miles of Texas in his Model T, visiting counties that had never before received a visit from a candidate for governor. Many of the roads he traveled were little more than unimproved wagon routes. Frugal to the core, Neff often camped, an activity he enjoyed, while on the campaign trail rather than staying in hotels or boarding houses.[10]

After his victory, Neff met staunch opposition to his progressive ideas regarding water conservation, efficiency in government, and prison reform. His idea to follow the national park model to create a state park system also fell on uninspired—if not deaf—ears at the state capitol. Despite opposition, in 1923 he was able to establish the State Parks Board to gather information and make recommendations on where to establish parks. He appointed David E. Colp (1867–1936) as board chair. Colp, who owned an auto dealership in San Antonio, had served as secretary of the Texas Highway Commission and was the point man for the Texas Good Roads Association. Neff and Colp saw a possible connection for promoting a state park system in conjunction with a state highway system.[11]

When Pat Neff was elected governor, his mother strongly supported many of his campaign ideas, and the concept of parks where people could gather and relax in nature especially intrigued her. In 1916 she donated six acres of her ranch along the Leon River to be enjoyed by the public as a place to picnic, congregate with family, or rest in the shade of a majestic pecan grove. Sadly, she died in 1921 during Neff's first year as governor. In her will, she requested that the six-acre grove be donated to the state for use as a park. Governor Neff proclaimed the donated land to be the first state park in Texas that was established solely for the purpose of outdoor recreation and not as a memorial park such as the San Jacinto Battlefield. Pat Neff later included an additional 250 acres to increase the size of the park. Mother Neff State Park officially opened in 1937.

The Texas Federation of Women's Clubs expanded its scope of work to include a Parks and Playgrounds Division under the federation's Depart-

Pat Neff with his mother, Isabella, date unknown. Courtesy Texas Collection at Baylor University, Waco Texas

ment of Conservation and Natural Resources. The other divisions under that department included Forestry and Wildlife Refuges; Natural Scenery; Billboard Restriction; Birds, Flowers, and Gardens; Highways and Memorial Tree Planting; Waterways; Historic Spots and Buildings; and Soils and Minerals. Each of the divisions was chaired by women who spanned the far reaches of the state, from Alpine to Port Arthur and Amarillo to Brownsville.

Despite not having modern conveniences such as cell phones, computers, highways, or Wanna Get Away Fares on Southwest Airlines, the women communicated frequently via letter, in person, and, occasionally, by telephone, which was becoming more prevalent even in rural communities. They waited patiently to hear from each other before taking action that required or warranted the approval of the collective. When necessary, they put their trust in a single individual to represent the federation on their subject matter of expertise.

Heeding the call of Governor Neff and the national call for "good roads," the federation called on its members to support national and state highway systems. In a 1923 article in the *Courier-Gazette* of McKinney, Texas, Decca Lamar West, chair of the Good Roads Division of the federation's Department of Conservation, made an appeal to rally support for a state highway amendment. A constitutional amendment was to go before voters to establish a state highway system that would construct and maintain roads and allow the state to levy a tax for that purpose. This was a radical change from the previous system of road construction and maintenance, which normally was controlled by individual counties. The Texas Federation of Women's Clubs noted the value in the amendment as twofold: first, it would allow for reliable transportation of rural schoolchildren; and second, it would lay the foundation for the creation of a state park system.[12]

The year 1923 also saw the passage of Senate Bill 73 by the Texas legislature, which established the first state parks board. Governor Neff appointed five members: David E. Colp from San Antonio, Phebe Kerrick Warner from Claude, Florence Martin from Dallas, Katie Owens Welder from Victoria, and Hobart Key Sr., a businessman from Marshall. All three women held leadership roles in the Texas Federation of Women's Clubs.[13]

Part of the board's early charge was to inventory the state to look for points of interest and encourage willing landowners to donate lands to be developed into roadside parks. This is where the federation excelled, as its local and regional officers had already been working to secure local parks and they had a valid inventory of points of interest and willing donors.

First state park board, 1924. *From left:* David E. Colp, Phebe K. Warner, Pat Neff, Florence Martin, Robert Hubbard (State Highway Commission chair), Katie Welder, and Hobart Key Sr. Courtesy Texas Parks and Wildlife Department, Austin, Texas

Governor Neff and David Colp envisioned securing smaller, roadside parks of fifty acres or less. Board member Phebe Warner and several of the club's leaders in West Texas had their eyes on bigger prizes and continually made the case for larger parks. The board made dozens of trips to identify points of interest and try to secure various properties, while also looking for suitable routes with the highway department to connect smaller towns with larger markets. Most of their travel was self-funded.[14]

By 1925, the Parks and Playgrounds Division of the Texas Federation of Women's Clubs was chaired by a capable and well-connected woman: Marian (Mamie) Rather Powell (1881–1974). Born in Huntsville, Mamie followed in her parents' footsteps to pursue a college education. She graduated from the University of Texas in 1902 and promptly returned to Huntsville to start her teaching career at the local high school. She was a skilled teacher who was appointed by the governor to the staff at Sam Houston Normal Institute in 1909. Per Texas state law, in 1913 she stepped down from her prized teaching duties at the college when she married a local up-and-coming attorney, Benjamin Harrison Powell III.[15]

Mamie did not go quietly into married life, never to be heard from again. With the support of her husband, she fought vigorously for women's rights, education, and improvements for economically disadvantaged families. She chaired the Walker County Suffrage Association, served on the board of the Red Cross, and was the first woman to serve on the Huntsville School Board.

Mamie and Ben welcomed their first son, Benjamin H. Powell Jr. in 1915 and their second son, Rawley Rather Powell, in 1918. Texas Governor William P. Hobby appointed Ben Powell III to the Twelfth District Court in 1919, and from 1920 to 1927 he served on the Commission of Appeals to the Texas Supreme Court. The dynamic couple moved to Austin in 1920 and immersed themselves in the energy of the capital city.

Tragedy struck in 1923 when their youngest son, Rawley, died at the age of five while undergoing a tonsillectomy. Rawley bled to death on the operating table. Bereft, the Powells created the Wildwood Sanctuary, a four-acre wooded park at the Oakwood Cemetery in Huntsville, in his honor.

Mamie Powell channeled her grief into one of the things she did best—public service. After all, service was in her blood. Her mother had been president of the Daughters of the American Revolution, and her father had served as mayor of Huntsville. Mamie was well respected in her own right as someone with a brilliant mind who was fiercely organized. She also knew how to leverage her connections and those of her parents and husband. The bridge clubs and garden parties held at the Powell home in Austin were filled with a who's-who list of Texas movers and shakers and their savvy spouses. These connections would serve her well as she stepped forward to embark on the journey to become the State Parks Board's secret weapon.

The letters retained by Nancy Powell Moore, Mamie and Ben's granddaughter, offer an account of exactly how integral Mamie Powell and the Texas Federation of Women's Clubs were in helping secure passage of the legislation that allowed the state to accept donated lands and create a state park system.

As chair of the Parks and Playgrounds Division of the Texas Federation of Women's Clubs, Mamie Powell established a strong relationship

Mamie Powell and her son Benjamin Powell IV, date unknown.
Courtesy Nancy Powell Moore

with David Colp and the other State Parks Board members. With the help of the vast membership of the Texas Federation of Women's Clubs, which numbered more than fifty thousand, the State Parks Board had been able to inventory and identify fifty potential parks with willing landowners who would deed the lands to the State of Texas. However, with Pat Neff out of office by this time, the Thirty-Ninth Texas Legislature had cooled its interest in accepting the lands or in creating a state park system. The State Parks Board needed a change agent to contextualize the relevance of parks, mobilize a grassroots movement, influence the inner circle of power, and do it all for free. Mamie Powell was just the person to do so.

At the annual Texas Federation of Women's Clubs convention in Dallas in November 1926, a resolution was adopted to petition the 40th Legislature of Texas to accept titles to certain park sites offered the State by patriotic citizens refused by the 39th Legislature. At the convention, the president of the Texas Federation of Women's Clubs, Florence Baldwin Fields of Haskell, and the other leaders expressed their confidence in Mamie Powell to champion the cause and represent the federation in all things regarding parks, even though Loretta Pearl Wells of Amarillo was chair of the Department of Conservation of Natural Resources. She was devoted but too far removed from the hub of political activity in Austin.

Another ally arrived at the state capitol in 1926 when Margie E. Neal (1875–1971) of Carthage clicked her heeled shoes up the granite steps and through the sturdy wooden doors of the capitol to be sworn in as the first woman senator in Texas. Margie Neal had already proved herself to be a savvy business owner and journalist when she owned the *East Texas Register* newspaper. She had also been an astute teacher and well-respected member of the Texas Federation of Women's Clubs. In fact, she served on many club committees alongside Florence Martin and Phebe Warner. Margie Neal had advocated for political equality in the suffrage movement and started her public service with an appointment to the State Teachers Colleges of Texas Board of Regents. While in Austin during the legislative session, she often visited the home of her longtime friend Mamie Powell and played bridge.[16]

With the Texas Federation of Women's Clubs on board and Senator Neal in the statehouse, David Colp and the State Parks Board now had the right team to encourage the legislature to accept the donated lands and create a state park system. In February 1927 Colp wrote to Mamie Powell to request that she and Mrs. E. A. Bellis of Fort Worth engage the Texas Federation of Women's Clubs in a letter-writing campaign to support Senate Concurrent Resolution 13 as introduced by Senator Neal to accept the properties. Colp suggested that a hundred letters would be a good number to sway the State Affairs Committee. Within weeks, more than four hundred letters poured in from influential women across the state.[17]

However, the first meeting before the State Affairs Committee did not go well, and all was almost lost. No shrinking violet, Mamie Powell did not hold back in her letter to David Colp to express her displeasure in how the matter was handled:

> Yesterday afternoon Miss Neal and I appeared before the State Affairs Committee in the interest of the resolution sent by you and introduced by her. Had you been here to give them actual information of the individual advantages of each site, I think the result would have been more favorable. As it was, neither Miss Neal nor I had visited each tract and could not assure them of particular merits. I was prepared to argue for acceptance alone unhampered by a program of maintenance, but when Senator Wirtz said he favored a State Park program but could not agree to accept the entire string of tracts, since he understood many were better fitted for county or town projects, I was silent except to say a very splendid group of citizens composing of Parks Board members having visited them were recommending them. I cannot understand why some members of the Parks Board did not consider the resolution of sufficient importance to come here to help meet the objections.[18]

Colp later wrote and apologized that he was detained on other business in the Rio Grande Valley.[19]

By mid-March, the tide had turned thanks to Mamie Powell's constant presence at the capitol, phone calls and letters to Texas Federation of Women's Clubs members, and the media. Senator Neal had a reputation for getting her bills passed and leaned on the remaining senators who needed a final push.

Even in the flurry of activity, Mamie maintained her busy family life with great care and affection. In a letter from Mamie to her mother, we have a window into the pleasures and demands of her personal life. She beautifully described her son and his friends gathered upstairs playing ping-pong while one of the younger visitors quietly read a book in her favorite corner of the house. The additional children were staying with

Mamie Powell in front of a painting of her grandchildren, circa 1950s. Courtesy Nancy Powell Moore

the Powells because her friend's youngest child had chicken pox, and her friend needed a safe place for her older children to stay.[20]

Mamie also noted that she had played bridge the previous day with her friends: Mrs. Bobbit (wife of the Speaker of the House), Mrs. Matson, Senator Neal, Mrs. Barry Miller (wife of the lieutenant governor), and

Mrs. Moody (wife of the governor). She wrote that during and after the game, they discussed the park resolution, which she described in detail. In the subsequent paragraph, she asked her mother to help her repair the lacework on her dress the next time she was in town. She concluded with how much she was enjoying her work on the park resolution and with the federation and admitted it was much more rewarding than cleaning the house or cooking.

Success occurred on March 17, 1927. Colp wrote Mrs. Bellis, "I suppose you had the pleasure of reading in the press this morning the final acceptance of twenty-three of our parks by the legislature, the John H. Kirby State Park at Fort Worth being among the number. We certainly owe you and Mrs. Powell a vote of thanks for your loyal cooperation."[21]

In the same letter, Colp added that while the parks had been accepted, they still needed to secure the appropriation of $50,000 and convict labor to get the parks ready for use. He requested that Mrs. Bellis be ready to organize another letter-writing campaign. While twenty-three parks were listed in the resolution, there were actually twenty-four properties donated to the state when the earlier gift from Isabella and Pat Neff was included.

Every major newspaper and many of the smaller ones picked up the story to celebrate the passage of Senator Neal's resolution to accept twenty-four parks totaling 1,858 acres across Texas. The *Austin American-Statesman* reported, "State Accepts 24 New Parks, Mrs. Ben Powell, Austin, Presses Resolution in Legislature." The article credited Senator Neal, Mamie Powell, and the Texas Federation of Women's Clubs for their skillful efforts in achieving passage of the resolution, which the governor approved on March 28, 1927.[22]

Securing the appropriations took a second burst of energy, and in a letter to Julia Owens of Navasota, Colp explained that the House was willing to support the expense; however, the Senate again was repelling the idea. He added, "I am going to Austin today for a conference with Mrs. Powell and shall be governed largely by Mrs. Powell's recommendations, believing that if she is given a free hand that she may be successful with this as she was in securing the acceptance of the parks."

Establishing a reliable method for funding state parks would take several more decades.[23]

Mamie Powell's legacy extends well beyond the establishment of the state park system, which now includes ninety-plus parks and wildlife management areas spanning over 640,000 acres of habitat and history. It lives on in each election that women participate in and the libraries, including the one in Austin, that she helped establish. It also lives on through her granddaughter Nancy Powell Moore.

In 1967 Nancy's parents, Ben H. Powell Jr. and Kitty King Powell, established a family foundation. In 1991 Nancy became the managing director of the Powell Foundation, where she carefully managed the charitable grants that benefited the communities of Harris, Travis, and Walker Counties, with a focus on education, conservation, social justice, and the arts.

In 1996 Nancy joined Ann Hamilton and David Todd in the Texas Environmental Grantmakers Group. She quickly saw the value and became a strong supporter of utilizing collective impact to achieve long-term system changes. After Nancy's children had grown and she had separated from her husband, she started looking for a place in the Hill Country that she could call her own. Following her grandmother's example, Nancy wanted to support conservation in her personal life with a piece of land that she could manage to encourage native species and native plants to thrive. She placed the majority of the property under conservation easements and consulted with Travis Audubon, the Lady Bird Johnson Wildflower Center, Texas Master Naturalists, and others to create baseline and ongoing studies of the plants and wildlife on the property. An early adopter of conservation easement, she shares her journey and lessons learned with other private landowners to encourage them to incorporate similar strategies on their properties as well.[24]

Each person involved in the establishment of the state park system has left an invaluable legacy, allowing every Texan or person who visits the Lone Star State to enjoy its natural wonders. The women who served on the first state parks board in 1924 are remarkable in their own right, and their individual efforts are worthy of a separate book. For the sake of space and time, the following is a snapshot of their contributions to improving the lives of Texans and conservation.

The Women of the First State Parks Board

Before serving two terms on the State Parks Board, Phebe Kerrick Warner (1866–1935) of Claude, Texas, was a journalist, writing a regular column in the Fort Worth Star-Telegram and periodic articles for other newspapers and journals across the state. She had a distinct journalistic style that captured the Panhandle and West Texas lifestyle and addressed many of the issues that communities faced, including the enfranchisement of women, hazards of industrial farming, county-based government, good roads, and, above all, improvement of the quality of rural life.[25]

Phebe Kerrick Warner, 1928. Courtesy Panhandle-Plains Historical Museum

Born May 2, 1866, in Belle Plain, Illinois, on her family's farm, Phebe earned a bachelor's degree from Wesleyan University. She taught at Illinois Women's University, where she established a natural science department before traveling to Texas, where she met her husband, William Arthur Warner. The couple were early settlers of the town of Claude in Armstrong County. The Panhandle is a vast, sparsely populated place even today, and it was even more so in the 1890s. Her husband was a doctor who frequently made house calls and provided insight into the remote home life of many of his female patients. Phebe also noticed that not all women in the community shared her love of the "oceanic plains." Many of them were isolated, alone all day with their children, and desperate for companionship or even conversation with another woman. To combat the loneliness, Phebe formed the Wednesday Afternoon Club and the Panhandle Pen Women's Club; later, the clubs were merged into the Armstrong County Federation of Women's Clubs.[26]

In 1917, the four-foot-eleven Phebe, otherwise known as the "Little Brown Wren of Texas," gave a passionate speech in Peoria, Illinois, that pushed her into the national spotlight as a champion for rural life issues, especially as they affected women and children. That same year, Governor William P. Hobby appointed Phebe to the National Congress of Farm Women. Phebe understood that good roads were key to improving transportation so that rural children could have the opportunity to attend school. She also astutely understood that quality roads increased business opportunities in towns, thus increasing taxes that supported the schools and the children they served. She used her pen and voice to advocate for the things she believed in, including the state park system.

In 1924, Phebe wrote about the need for parks in her column for the *Western Weekly*: "Conservation. Recreation. Education. These are the underlying motives in the establishment of a chain of parks throughout the state of Texas. There is no other state in the Union so rich in historical interest and scenic beauty. But who knows it?"[27]

She continued, "But men and women of Texas, you may not know it yet but there is one of the greatest civic movements afoot right now in Texas that this nation has ever known. . . . I believe the plan for a chain of State Parks for Texas is going to become not only national but international in a few years, and the day is not far distant when the beauty and the history of this great land of ours will be linked together by the finest system of state and international highways in the world and every citizen of our nation as well as Texas will be within a few hours and a few cents of some beautiful creation of nature."

Palo Duro Canyon was suggested as a national forest reserve as early as 1908 by US Congressman John H. Stephens of Wichita Falls. However, a handful of people had been advocating since the 1890s for the canyon to become a park of some kind. Phebe wrote about the topic in the *Western Weekly* on June 28, 1925, "Beauty might be considered one of the state's natural resources, the same as oil or cotton. If so, then why is it not the duty of any state to develop its scenic resources for the good they will do the world and the value they may be to their own people? There is in the heart of the great Texas Panhandle one of nature's rarest scenic gifts."[28]

She added that the people of the Panhandle needed to petition the legislature to establish a highway to the canyon so that it would rank high

Phebe Kerrick Warner standing to the right of Pat Neff touring Palo Duro Canyon with members of a local women's club, date unknown. Courtesy Panhandle-Plains Historical Museum

on the list for possible parks. In the same article, she touched on a theme that she frequently mentioned: Texas was losing money to Colorado because Texans were deciding to travel there to admire the state's natural beauty instead of enjoying that of their own state.

Despite Pat Neff and David Colp's idea to establish small parks every fifty to one hundred miles along highways, Phebe held true to the idea that large parks had a purpose and place in the system. Colp later got on board and tried throughout the 1930s to bring Palo Duro into the system as others tried to get it designated as a national park. Also during this time, President Herbert Hoover appointed Phebe to the national Home Builders Committee, and she later ran unsuccessfully for the United States Congress.

Always advocating for Palo Duro to become a park, Phebe wrote the following in the *Southwest Plainsman and Panhandle Weekly*: "The God of Nature is a most kind God. One who believes in plenty of sunshine and beauty, rest, recreation and pleasure for us all. A divine Artist whose private gallery is the world." Like many of her naturalist friends, Phebe did not believe in the beautification of nature. She subscribed to the view that nature was beauty, each stone, each river, each living thing placed with divine purpose.[29]

In 1933, in the heart of the Great Depression, the windfall the park campaigners needed arrived thanks, in part, to Vice President John Nance Garner of Uvalde and Reconstruction Finance Corporation Director Jesse H. Jones of Houston. These men advocated for Texas to receive a steady stream of financial aid from the New Deal for the construction of roads and funding from the Civilian Conservation Corps and Works Progress Administration to build the dams, roads, cabins, dining halls, and trails for the new state park system. Ultimately, it was Colp who secured a Reconstruction Finance Corporation loan to purchase the privately owned Palo Duro Canyon.

Palo Duro Canyon State Park opened on July 4, 1934. More than forty thousand visitors enjoyed the natural wonders of the park in its first year. In that same inaugural month, Phebe's husband, William, passed away, with Phebe following just a few months later from complications with pneumonia. On May 18, 1935, the *Fort Worth Star-Telegram* reported, "Perhaps more than any other person, Mrs. Warner may be described as the champion of Palo Duro Canyon Park in the Panhandle. The existence of that magnificent park and playground today is one of the lasting monuments to her memory."[30] Her papers and memorabilia can be found at the Panhandle-Plains Museum of Texas in Canyon.

Florence George Martin (1875–1958), more often referred to as Mrs. W. C. Martin, served as vice chair of the State Parks Board until the 1930s. She was also well known for her efforts to improve the lives of tenant farmers and for introducing a revolutionary model farming system in Darco, Texas. She wanted to prove that tenant farmers were not "shiftless," as many believed them to be, and demonstrate that they could become productive farmers if given the right tools, training, and amenities of a stable community complete with access to schools, churches, libraries, medical services, and businesses. Her model even ensured that farmers had access to a community phone so that they could call a doctor, the fire department, or police if needed. Access to the last was a con-

Florence Martin posing in a Cadillac Model E automobile, 1906. Courtesy F. Ed. Spooner, Lazamick Collection, National Automotive History Collection, Detroit Public Library

stant struggle for many rural families. She self-funded the program on land that she and her husband owned.[31]

Florence married William Clifton Martin, a businessman and rancher who seemed content to support Florence and let her do as she pleased—and what pleased her was ranching, educating, volunteering, and writing. William died in 1927, and Florence continued managing their lands until her death in 1958. In the 1920s, lignite was discovered on a portion of their properties near Darco, which further gave her the financial means to pursue the causes she deemed worthy.[32]

She chaired the Country Rural Life Division of the Texas Federation of Women's Clubs for many years, and was the editor of the *Texas Federation News*. She was the first woman chair on the National Farm Bureau and spent a great deal of time writing and speaking on the topic of rural life. Governor Pat Neff thought highly of Florence and appointed her to be the only woman on the Penitentiary Relocation Board as it wrestled with the issues of prison farms, prison labor, and reform.

Above all, Florence loved Texas and believed that each person should "know their state first." In an article she wrote for the *Fort Worth Star-Telegram* on July 13, 1924, she articulated her feelings on the topic: "It should be our first duty, to know our own state, and our first trips should be made there. When, after that ground has been covered and the family has been educated it is time to take a post graduate course, and go on into the neighboring states, the great transcontinental highways and our state and national parks have also made traveling by motor a real recreation."[33]

In the same 1924 article, she wrote about the "Glacier to Gulf Highway," which was a concept to create a major highway system that ran from Glacier National Park in Montana to the Gulf of Mexico, with state and national parks located on or near the highway every hundred miles. Sadly, that amazing idea did not fully materialize. Always the advocate for lower- and middle-income families, Florence saw motor travel to parks as an inexpensive means to travel and make lasting memories.

Florence also had farms and businesses in and around Marshall, Texas, which motivated her to advocate for designating Caddo Lake a state park. In 1924, members of the State Parks Board and interested parties from Louisiana visited Caddo Lake as they considered it for inclusion on the list. In 1928, in an article in the *Fort Worth Star-Telegram*, Florence stated that three larger parks were being considered that would be approximately ten thousand acres each: Caddo Lake, Davis Mountains, and Palo Duro Canyon.[34]

In 1934, heeding the call to donate land around the lake, men such as Thomas Jefferson Taylor II gifted a portion of their land to be included in the 385 acres first set aside for Caddo Lake State Park. T. J. Taylor was the father of Lady Bird Johnson and her uncle, W. T. Taylor, held the first concessions contracts at Caddo Lake and Palo Duro Canyon. Caddo Lake opened in 1934 after the Civilian Conservation Corps constructed cabins, a dining hall, roads, and other infrastructure.

After serving on the first state parks board, Florence continued to support rural life improvement projects, and during World War II she served as president of the Women's Auxiliary for the Red Cross.

Katie Owens Welder, circa 1950s. Courtesy Victoria Regional History Center Special Collections, University of Houston, Victoria, Texas

Katie Owens Welder (1864–1958) was born in Victoria County, Texas, during the Civil War. She married James Francis Welder, a prominent rancher and banker, in 1889 and the couple had five children. In 1912 she formed the Mother's Club to support St. Joseph's School in Victoria and continued to serve that club for more than thirty years.

In 1922, she was named chair of the Parks and Playgrounds Division of the Texas Federation of Women's Clubs. She attended the Second National Conference on State Parks with two other Federation representatives. The conference was sponsored by the National Park Service at Palisades Interstate Park, New York. While there she stated, "I feel that I cannot let this conference close without a word from Texas. I want to say that I travelled four days to come to this conference with no other objective but to hear and observe what these wonderful achievements have accomplished in the way of parks and playgrounds. I am sure that you all feel Texas is asleep, but she isn't; she is very much awake and I am going back full of ideas, which I hope someday to put into action."[35]

Katie went on to serve on the State Parks Board under four governors. She also held various leadership roles in the Texas Federation of Women's Clubs, including chair of the Permanent Headquarters Committee. Building a permanent headquarters, known as the Mansion, incurred

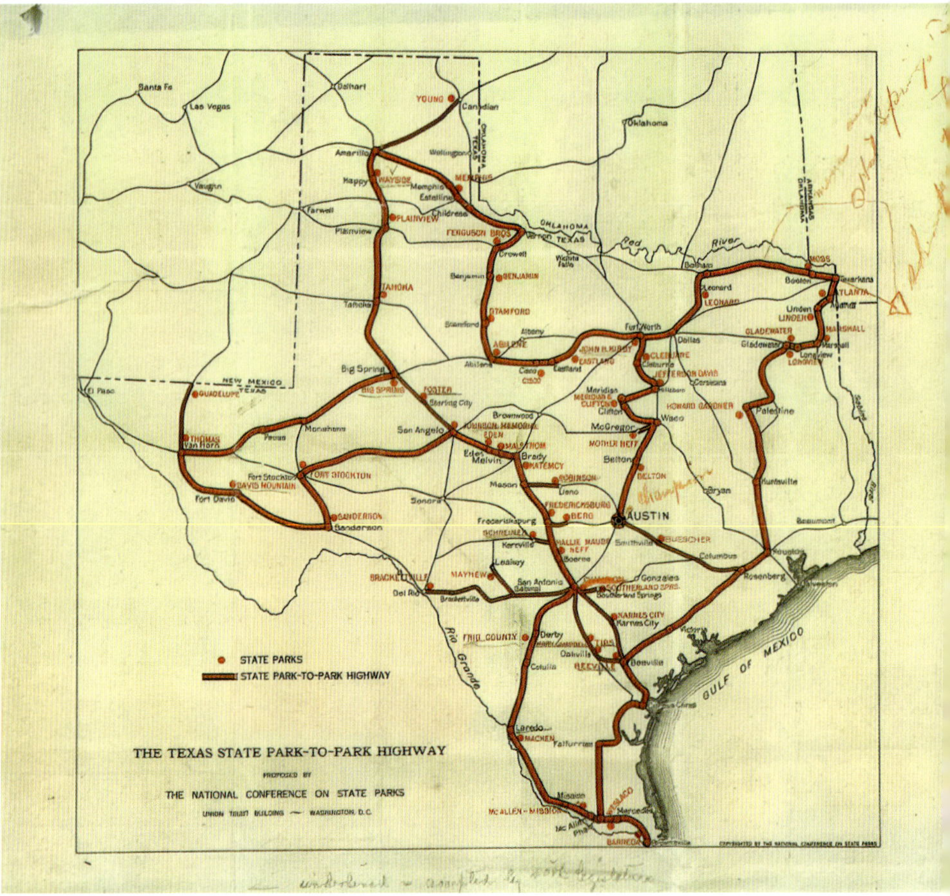

The Texas State Park-to-Park Highway map, 1927. The map reflects the proposed parks; a note included with the map indicates that only the parks underlined were accepted into the system. Courtesy Nancy Moore Powell

tremendous debt as funding dried up during the Depression. Katie never gave up trying to find a solution. In 1939, when all seemed lost, fellow club member and philanthropist Clara Driscoll stepped forward to pay off the $92,000 in debt to finish construction and create a clean slate for the federation to move forward with its mission. and enjoy its new clubhouse in Austin.

Closer to home, Katie was always active with her charity work for the Catholic Church and received honors three times from the pope in recognition of her efforts. Katie was also a charter member of the Gonzales

Warm Springs Foundation for Crippled Children, which utilized the mineral-rich springs in rehabilitation efforts for children with physical impairments. She also served the Bronte Club for more than fifty years. The Bronte Club of Victoria, founded in 1855, is the oldest women's literary club in the state. The club sponsored a lecture by First Lady Eleanore Roosevelt in 1940; the event was attended by twenty-five hundred people.[36]

Katie was widowed in 1931 and lost both her sons by 1941. After the death of her sons, she took over all aspects of their vast cattle ranches and oil and gas operations. She passed away at her home in Victoria in 1958 at the age of 94 years old.

After the service of these first founding women of Texas state parks, it would take sixty-four years before a woman would again serve in a leadership role on the governing body for the Texas Parks and Wildlife Department. Beatrice (Bea) Pickens, wife of oilman T. Boone Pickens, was appointed to the Texas Parks and Wildlife Commission in 1987 by Governor Bill Clements. At the time, Bea was the director of the National Fish and Wildlife Foundation and an avid hunter who understood the need for good habitat to support fish and game. Her appointment opened the door for a succession of women to serve on the commission, including Terry Hershey and Mickey Burleson, both appointed by Governor Ann Richards. It would take another ten-plus years before one of the female commissioners became chair. In 1999 Katherine Armstrong was appointed to the commission by Governor George W. Bush; she became the first female chair in 2001. The eleven-member commission has never had more than two women serving at the same time despite an effort to change that dynamic for all state commissions in 2017.

Each of these women harnessed her particular abilities and interests to significantly contribute to forming and sustaining the Texas state park system. They did not just lay the foundation, they also gathered their friends and families to clear the grounds, mix and pour the cement, and host a luncheon to celebrate their collective achievements. They approached each action with an open mind devoid of self-aggrandizement. They juggled the demands of family life, the challenges of expanding social norms, and the doubts of the established power structure, all to serve the greater good of their fellow citizens and leave a legacy that their children and all future generations would enjoy.

and Black-capped Vireo nested. The report also identified the creek systems as areas of critical habitat where a matrix of conservation lands should be set aside to mitigate flooding and provide habitat, with additional lands set aside for recreation parks.[37]

That same year, a group of citizens in the Eanes Independent School District and the city of West Lake Hills also became concerned about the pressures and strains of new development along the coming highway. Under the leadership of Nancy Bowman, Now or Never was founded with the mission to "help preserve the natural resources, tranquilities and amenities of the area." In its first years of action, the group worked with and highlighted developers who took the environment into consideration in their projects.[38] This was a decisive strategy, as it gained the group favor with the development community. Around this same time, Holye Osborn was preparing a master plan for the City of West Lake Hills. Holye made the suggestion to two Now or Never members, Martha Hudson and Janet Poage, that the land along Bee Creek be made a state park.[39]

The idea was further driven home when architect and conservationist Sinclair Black gave a presentation to Now or Never titled "Waterways of Travis County." He stated that the highway would devastate the environmentally sensitive Barton, Eanes, and Bee Creeks. Sinclair's presentation was an outgrowth of the Horizons '76 Bicentennial plan to preserve the creek ways in Austin and Travis County. The plan called for park and trail systems along the creeks to connect communities, neighborhoods, and schools to places of cultural significance and to be used for recreation. Martha Hudson, president of Now or Never at the time, suggested that a committee be formed to pursue the creation of a park. Janet Poage stepped up to chair the committee.

In 1975 Now or Never adopted the park idea as a Bicentennial project.[40] Janet and Martha recruited five of their friends to the committee, and the journey to establish the Wild Basin Preserve began. The committee included Martha and Janet, plus Willa Mae Hardesty, Margaret Hessin, Flo Prymer Macklin, Lucille Stegman, and Esther (Flora) Wargo McCormick.

In a 2019 interview, Martha Hudson and Flo Prymer Macklin laughed as they concurred that Janet Poage was the leader of the seven, and everyone else followed along because whatever Janet did was exciting.[41] They went on to describe one of the first times that the seven women ventured into

Janet Poage, Sue Johnson, and unnamed man holding "Wild Basin Wants You" bumper sticker, date unknown. Courtesy St. Edward's University Archives and Special Collections

the Wild Basin to get a sense of what they were saving. They walked up the creek bed from Westlake Drive equipped with only a picnic basket and bottle of wine. They didn't have a map, nor was there a trail for them to follow. Martha said that day was the perfect analogy for how unprepared, but totally willing, they were to embark on this complicated journey.[42]

Janet Poage (1919–1993) was the niece of renowned suffragette and activist Jane Yelvington McCallum (1877–1957) of Austin. After fighting for women's right to vote, Jane went on to fight for reform of child labor laws, improved conditions in prisons, and women's health issues in Texas. Texas legislators mocked Jane and the group of women she assembled by calling them the "Petticoat Lobby." Jane was later named the first woman secretary of state by Governor Dan Moody and is credited with preserving the Texas Declaration of Independence.[43]

As a youth, Janet resided for a short time with her aunt and uncle. In a March 2, 1986, interview printed in the *Austin American-Statesman*, Janet

Martha Hudson at a fundraiser at the Zilker Clubhouse, date unknown. Courtesy St. Edward's University Archives and Special Collections

credited her aunt as having a powerful influence on her life, especially in the realm of perseverance for a worthy cause. However, Janet's causes had to wait as she traveled the world with her husband, Oren Poage, who was a member of the US Air Force. Oren rose to the rank of colonel before retiring in the 1970s to the hills west of Austin.[44]

Janet was by no means idle during those years as she raised their five children and volunteered at the Red Cross. Before her involvement with the Wild Basin, she served as chair of the Zoning and Planning Commission in West Lake Hills. In 1992, just before she died, Janet wrote an unpublished book recording the journey to save the Wild Basin. The book has one of the best titles and perfectly captures Janet's wit: "Snowball in Hell, or How Seven Little Old Ladies in Tennis Shoes Whipped the Establishment and Created an Urban Nature Preserve in Austin, Texas." In the manuscript, Janet described the other women committed to the cause.[45]

In "Snowball in Hell," Janet described her coconspirator, Martha Hudson, as "a home-grown Texan, down-to-earth practical, conservative with a good sense of humor and a great tolerance for the foibles of her fellow human beings. She [Martha] once said that she felt one of her main contributions to the Wild Basin had been to 'hold Janet's feet to keep her from flying off into the stratosphere from time to time.'"

After graduating from North Texas State University, Martha became a teacher for several years before her marriage to Ronald Hudson. The couple raised three sons in the Eanes school district, where Martha volunteered on the parent-teacher association and at school libraries. She was elected to the Eanes Independent School District as a board member, with one term as president in the 1980s.

Janet was not the only member of the committee who had roots anchored in the suffrage movement. Margaret Eckerea Hessin (1916–2005) is reported to have attended a suffragette meeting at the age of three. She earned her psychiatric nursing degree from Samuel Merritt College in California before marrying Arthur Hessin. Janet described Margaret as someone who "had considerable women's club experience, was a superb chair for a meeting who would stick with Robert's rules of order and always completed an agenda on time."[46] In 2001 the United Way of Central Texas honored her with the Volunteer of the Year Award.

With Janet as their leader, the women wisely started their campaign by reaching out to top leadership, including US Congressman Jake Pickle. He and his staff directed them toward the Land and Water Conservation Fund administered by the US Department of the Interior. But they would need a government entity to apply for and manage the grant. They sought willing allies on the Austin City Council and Travis County Commissioners Court. To form the preserve, it was determined that they would need about $700,000 to purchase a total of about three hundred acres.

In 1976 conservationist Ned Fritz of The Nature Conservancy formally submitted an amendment to "The Listing of the Thirty Most Significant Urban Natural Areas in Texas as Presented in the Texas Outdoor Recreation Plan." The amendment recommended that both sides of Bee Creek, approximately three hundred acres, be conserved. The Texas Parks and Wildlife Department was approached about making the acreage a state park. The agency determined the area to be too small, plus the highway would be an issue. Instead, the women convinced the City of Austin and Travis County to pledge $175,000 each to secure the matching Land and Water Conservation Funds. Everything seemed to be moving smoothly.

Without warning, the city pulled its commitment to the Wild Basin. The blow almost derailed the entire project. In an interview, Martha Hudson and Flo Macklin shared that the rebuff just made Janet Poage mad and want it even more. They also shared that Janet was committed to making sure her husband always had dinner ready by six. While her husband was supportive of Janet volunteering, he did not want those activities brought home. She often had to use the phone at one of the other members' homes to conduct the business of saving the Wild Basin.[47]

An important ally entered the picture. Ann Richards had served on the West Lake Zoning and Planning Commission with Janet Poage and was elected the first female commissioner in Travis County. The Now or Never committee had already secured the support of Janey Briscoe, wife of Governor Dolph Briscoe, who served between 1973 and 1979. They also raised an army of botanists, geologists, professors, and other conservationists to assist with building a science-based case around why this piece of Hill Country was worth saving.

Without the city's commitment the county also had to drop its application to the Land and Water Conservation Fund. County Commissioner Ann Richards suggested that if Now or Never could raise the $175,000, the county could apply for a Bureau of Outdoor Recreation grant and commit to matching the funds, but it had to be done quickly. The seven women decided they would try and hope for the best.

Once they started into the process of raising money, they engaged everyone they possibly could and were open to just about any idea—from having kids stand outside Rylander's Grocery Store to hosting a fancy fashion show fundraisers. Third graders sold cookies and pies. Churches held rummage sales. A popular roller-skating rink in South Austin held an event that drew in the US Department of Labor's Ray Marshall and his family, who were residents of Austin. T-shirts were sold that said "Save the Wild Basin." The community was intrigued, and the women raised $80,000 in three months.

They even held a Christmas in June celebration. In 1977 the *Austin American-Statesman* ran announcements throughout the fundraiser, publicizing the amount the committee still needed to raise in order to reach their goal and instructing readers where to send money to help "Save the Wild Basin." The ladies even gained permission (thanks to Janey Briscoe, the governor's wife) to put a Christmas tree on the grounds of the capitol. If they had an idea that seemed fun, doable, and worth the time, they did it. And the media loved them for it.

On New Year's Day 1977, a fundraiser was held in the giant Quonset hut on the banks of Town Lake known as the City Coliseum. The event boasted seven bands; dozens of booths filled with jewelry, macramé,

Wild Basin fashion show fundraiser featuring designers Angiona and Victor Costa, 1981, *Austin Citizen*. Courtesy St. Edward's University Archives and Special Collections

pottery, and food; and even a "budding artist" booth for kids. More than five thousand people attended the event (including me and my family). Flo faithfully volunteered for each big event.

Flo Prymer Macklin (1928–2021) grew up in Ohio and earned her degree in retail merchandising for women from the Ohio State University. She traveled the country conducting market research for Procter and Gamble before marrying air force officer John Macklin. In 1970 the Macklins and their three children settled in the suburb of Rollingwood when John accepted his final assignment at Bergstrom Air Force Base in Austin. Flo volunteered for the next forty-plus years with organizations such as No Zoo in Zilker, Save Our Springs, the Republican Women's Club, the Rollingwood History Project, the West Lake Hills Presbyterian Church, and the Westbank Library. She was nationally recognized for volunteering for forty-three years with Meals on Wheels, and in 2015 the Westbank Library named its community room in her honor.[48] In the 2019 interview, Flo

Flo Macklin and Flora McCormick making cotton candy at the annual Canterbury Faire fundraiser, date unknown. Courtesy St. Edward's University Archives and Special Collections

described herself as a foot solider, willing to do what was asked for the Wild Basin project. She loved nature and especially wanted a place where children could learn and explore the outdoors.[49]

Every dollar raised was carefully counted and recorded by Lucille and Willa Mae. While there were multiple large donations from businesses and citizens, the average gift was around twenty dollars. Willa Mae was known for saying they raised the money by selling one glass of lemonade or slice of pie at a time.

Lucille A. Brightbill Stegman (1924–2007) graduated from Oberlin School of Commerce and was an executive secretary for many years, serving as president of the Austin Association of Executive Secretaries. Before moving to Austin in 1961 with her husband, Harold, she had been a flight attendant for United Airlines during World War II. In addition to her work with the Wild Basin, Lucille fought, and won, to keep billboards and other lighted signs out of the city of West Lake Hills and away from Highways 2244 (Bee Caves Road) and Loop 360.[50]

Willa Mae Boozer Hardesty (1931–2023) earned her degree in home economics and enjoyed a career as a teacher for several years. Her husband, Boyd Hardesty, was a chemistry professor at the University of Texas; they had three children. She was active in the League of Women Voters and Save Our Schools before joining what she called the "Wild Bunch." Janet Poage described Willa Mae as "a woman who could fix anything, from her car to the plumbing in their home to whatever problem needed a solution."[51]

In addition to raising money, the women had set a goal of securing some of the land via donation from the landowners. In 1977, they succeeded at both. Landowner and developer Sid Jagger sold Now or Never thirty-five acres well under market value. The donation of land opened the door for additional tracts to be purchased in a similar way. With twenty landowners, negotiations took time.[52]

Despite their success in raising the funds and securing the first property—with more on the way—the women's application was turned down by the Texas Parks and Wildlife Department. The department was the gatekeeper of applications that were forwarded to federal agencies for land acquisition and conservation. Janet Poage wrote about that critical time in "Snowball in Hell":

> The ladies rolled out the cannon again. Governor Briscoe was in Washington and put in a word to the Interior. Congressman Jake Pickle and Senator Lloyd Bentsen and John Tower agreed to help. Local testimonials were profuse. 1978 was an election year for Senator Tower, a Republican. He was being challenged by Bob Krueger, a Democrat, both Wild Basin supporters. Ann [Richards] as a practicing Democrat, saw an opportunity to help Bob Krueger's candidacy. When it became clear that Interior would award Wild Basin the requested contingency funding [a separate request after being denied the other grant] Ann suggested that Bob Krueger be given credit for getting the money, and that he make the announcement of the award to the press. The ladies had always sought non-partisan support for the Wild Basin, and they didn't care who made the announcement, so long as the funding was forthcoming. Krueger was upstaged, however, when the Interior Secretary, Cecil Andrus,

announced that he would come to Austin, tour the Wild Basin, review the application and make the announcement himself.

Secretary Andrus was also in town to present the ninth annual Lady Bird Johnson Highway Beautification award. Media excitement was through the proverbial roof. In a last-minute play, Senator Tower arrived early at the capitol to hand out flyers stating that he had secured the funding. This created more media frenzy. In the end, the contingency funds were granted.[53]

In 1979 Wild Basin Wilderness, Inc., was formed to manage the flow of money and make the land purchases. During that year, the organization held a fundraiser that included an auction to pay for brokerage fees, surveys, and some land acquisition beyond what the federal dollars could support. The list of auction items included a copy of the screenplay for *All the President's Men*, which was signed by actor Robert Redford. Other items included a lot in a new subdivision in Dripping Springs and artwork from several local artists. The celebrity items, of course, caught the attention of the newspapers, especially of columnist Mike Kelly, who had been reporting on the Wild Basin's progress for several years.

One of the other celebrity items was a pillow that Elvis had used when he performed in Austin. The pillowcase was a gift from an eclectic man who resided in a local hotel. Flora rushed to the hotel to pick it up. Wild Basin Wilderness, Inc., worked out a deal to auction it off on KLBJ Radio. According to Janet, a Mrs. Vernon Smith of Elgin was getting her hair done at the beauty salon when she heard the auction on the radio and called in to outbid everyone. (In those days listeners called the station via landline phones to place their bids. Most callers got a busy signal.)

As the 1970s gave way to the 1980s, the era of consumerism took hold of the country; Austin was not excluded. Properties were selling so fast that it became difficult to know who the current owner was. The inflated prices slowed purchase of the final parcels of land. One of the last big pieces of land was added in 1989. Developer Beth Robertson, City of Austin Mayor Carol McClellan, and County Commissioner Ann Richards reviewed the master plan of Davenport Ranch residential development in order to set aside over 100 acres to complete the preserve.

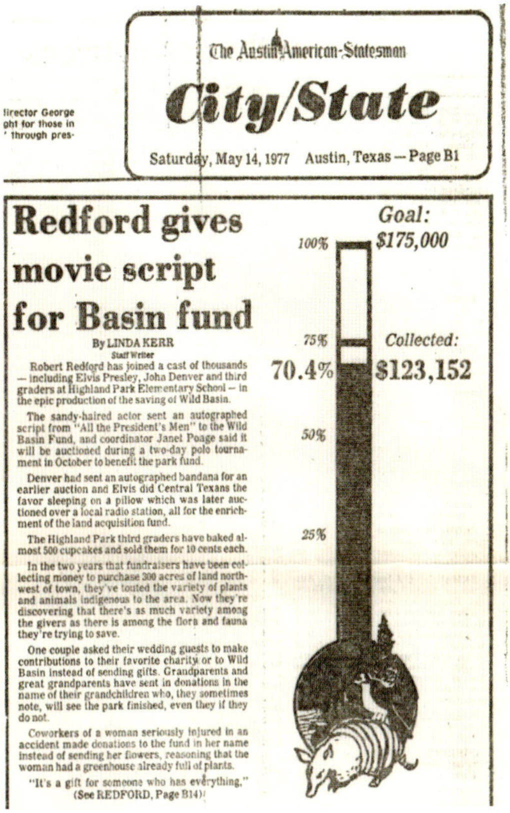

lirector George
ght for those in
' through pres-

The Austin American-Statesman

City/State

Saturday, May 14, 1977 Austin, Texas — Page B1

Redford gives movie script for Basin fund

By LINDA KERR
Staff Writer

Robert Redford has joined a cast of thousands — including Elvis Presley, John Denver and third graders at Highland Park Elementary School — in the epic production of the saving of Wild Basin.

The sandy-haired actor sent an autographed script from "All the President's Men" to the Wild Basin Fund, and coordinator Janet Poage said it will be auctioned during a two-day polo tournament in October to benefit the park fund.

Denver had sent an autographed bandana for an earlier auction and Elvis did Central Texans the favor sleeping on a pillow which was later auctioned over a local radio station, all for the enrichment of the land acquisition fund.

The Highland Park third graders have baked almost 500 cupcakes and sold them for 10 cents each.

In the two years that fundraisers have been collecting money to purchase 300 acres of land northwest of town, they've touted the variety of plants and animals indigenous to the area. Now they're discovering that there's as much variety among the givers as there is among the flora and fauna they're trying to save.

One couple asked their wedding guests to make contributions to their favorite charity or to Wild Basin instead of sending gifts. Grandparents and great grandparents have sent in donations in the name of their grandchildren who, they sometimes note, will see the park finished, even they if they do not.

Coworkers of a woman seriously injured in an accident made donations to the fund in her name instead of sending her flowers, reasoning that the woman had a greenhouse already full of plants.

"It's a gift for someone who has everything."

(See REDFORD, Page B14)

Austin American-Statesman article announcing Robert Redford's gift and the donation thermometer tracking funds raised, 1977. Courtesy St. Edward's University Archives and Special Collections

With the funds now being managed by Wild Basin Wilderness, Inc., and the county, Now or Never dissolved, and several of the founding women moved on to focus on new issues affecting the community. Even though Janet was the most public and persuasive member of the group, Esther (Flora) Wargo McCormick earns the honor of giving service to the Wild Basin the longest. Janet stated, "Flora was respected among her Wild Basin peers for her total honesty and her absolute adherence to the truth. Ask her for her opinion and she would give it to you, unvarnished. When she signed on for a project, you could be sure she would give it 150 percent."[54] Flora earned a registered nurse license and became an instructor in the Medical Surgical Nursing Department at St. Luke's Hospital in Bethlehem, Pennsylvania.

Over the years, Flora was a tour guide for the Wild Basin and became the head of the Environmental Education Committee. She was widely

Poster for "Keep Austin Wild—Christopher Cross Party" to benefit the Wild Basin Preserve, 1982. Courtesy St. Edward's University Archives and Special Collections

viewed as the most popular tour guide because of her extensive knowledge of the basin's flora, fauna, and geology and her ability to impart that knowledge to visitors and schoolchildren. She also helped maintain trails and write the outdoor education training guides.[55]

The legacy of the women of the Wild Basin extends well beyond the 227 acres now preserved for all to enjoy. Today, the Wild Basin Creative Research Center is managed by St. Edward's University and is used as a living classroom. Since the 1980s, it has hosted thousands of schoolchildren and families for nature-based educational programs. The Wild Basin remains a slice of the Hill Country that is rapidly disappearing as Austin continues to be one of the fastest-growing cities in the United States.[56]

Barabra Dugelby, the director of the center, has helped expand the education and research programs. With assistance from volunteer, naturalist, and former Wild Basin Preserve staff member Terri Seigenthaler, the center began the process of archiving and digitizing the articles and photos of the foundational women. They are capturing the stories of many of the other volunteers, board members, and staff. The collection is now stored in the Mundy Library at St. Edward's University Special Collections in Austin.

A CHAMPION OF ART, MUSIC, HISTORY, AND NATURE

Jane Dunn Sibley (1924–2019)

Jane Dunn Sibley came from a long line of determined, independent women. One of the family legends describes her great-great-grandmother reading William Shakespeare to her five children as their cabin in Mason County, Texas, was surrounded by Comanches. All survived. Later her grandmother and mother, along with the rest of the family, traveled to West Texas to settle in the frontier towns of Alpine and, later, Fort Stockton.

Jane describes her childhood in Fort Stockton as pleasant. She swam in Comanche Springs and participated in water ballets with her friends. In 1941, just a few months before the US entered World War II, Jane moved to Austin to attend the University of Texas. It was in Austin that she found the mental stimulation she had been craving, and her love affair with the capital city began.

Jane Sibley, 1960s. Courtesy Sibley family

After college, she returned home to Fort Stockton and eventually met and married D. J. Sibley Jr., her dashing partner for fifty-four years. D. J. Sibley had also grown up in Fort Stockton and was a friend of the family. He returned home from the South Pacific theater of World War II a decorated war veteran. Unfortunately, while serving in the medical corps, he contracted tuberculosis and suffered from it the remainder of his life. Despite his illness, he had a bright and talented mind and a zest for life. Like Jane, he had a love of art, history, and nature and is noted for his support of the University of Texas Natural History Museum, Bat Conservation International, and Chihuahuan Research Institute, among other interests. Most importantly, he fanned the flames of Jane's passion for the arts and nature and supported her dynamic adventures.[57]

It is hard to recap Jane's impressive life as she helped transform Austin from a capital city with a university into a place where art and music flourished. In 1964 she joined Laguna Gloria Art Museum (now the Contemporary Austin) and turned it into the epicenter for art in Austin. She elevated the struggling museum to financial prosperity through shrewd fundraising skills and fiscal management. In 1967 she joined the board of the failing Austin Symphony Orchestra and brought it from the brink of bankruptcy to financial stability. She served as its president and champion for twenty-five years. Jane saw a clear need to entrench both organizations in the community through events, outreach, and media. She also championed both the Long Center for Performing Arts and Symphony Square projects in Austin. Whatever she was doing was cool, and that made others want to get involved. In 2000 Symphony Square was renamed in her honor.[58]

In 1968 the US Army Corps of Engineers was supervising the construction of Amistad Dam along the Rio Grande north of Del Rio, Texas. The National Park Service was working feverishly to remove more than a million artifacts and document the rock art of the soon-to-be-flooded area. Vandals and thieves were prowling onto private lands to remove artifacts and deface rock art sites. Thanks to Ellen Schultz Quillin and Emma Gutzeit, the Witte Museum in San Antonio had already started documenting and securing artifacts decades earlier.

Texas Governor John Connally was gearing up for the 1968 World's Fair at the HemisFair in San Antonio, with the idea of celebrating the rich diversity of Texas' people, ingenuity, and natural wonders. While his team was preparing, the unique cultural sights of the Pecos and Rio Grande began generating interest among museums, conservationists, and the governor's office. One of the ideas the governor supported was to remove the rock art located along the Pecos River and reassemble it at the fair. Engineers from Brown & Root researched the idea and informed the governor that removing the paintings intact would be astronomically expensive. Most importantly, the move would destroy the art. Brown & Root suggested that it would be more cost effective and better reflect the culture to enjoy the art in a parklike setting at the location they were painted. At that point, the site made it onto the Texas Parks and Wildlife Department's long list of places to consider for park status. Thanks to Governor Connally, who pushed the legislature to approve the first-ever bond package of $75 million to purchase new parks and historic sites, the idea seemed feasible.

Jane Sibley observing the rock art on a trip to the Lake Amistad area, 1981. Photo by Patrick Horsborough. Courtesy Sibley Family

The state's archeologist, Curtis Tunnel, invited his friends D. J. and Jane Sibley to tour the area to share the wonders of the rock art before it was submerged underwater. Early humans entered the region approximately twelve thousand years ago during the Pleistocene Ice Age. It was not until four thousand years ago, during the Middle Archaic period, that the first artists adorned the rock overhangs with paintings.

The Sibleys were longtime advocates for historic preservation and regional culturalism and understood the significance of the pictographs. Jane was instantly enchanted with the rock art and the land. However, she was not able to take action, as D. J. suffered a relapse of tuberculosis in 1968, which, along with raising her three children and working in Austin, required her full attention.

While attending a fundraising event in 1970, Jane had a conversation with her archeologist friend Solveig Turpin (1936–2020) about the remaining rock art along the Pecos. Solveig provided the push Jane needed as she shared, "Oh, Jane, the vandalism is just awful." Jane and Solveig quickly recruited their longtime friend Rose Mary Whitehead Jones (1925–2015). Rose Mary came from a successful Boer goat and cattle ranching family from Sonora and Del Rio and was already involved in historic preservation in Val Verde County. The three women went to work to protect the historic site. Jane and her team invited Governor Preston Smith to tour Panther Canyon. Admittedly, the governor knew or cared little about the rock art, but he appreciated the cultural value—and Jane's tenacity—so he appointed her to the Texas Historical Commission later that year.[59]

Unlike the rapidly paced acquisition of Enchanted Rock State Natural Area that occurred around the same time, it would take more than twelve years to convince the state to preserve, purchase, and open Seminole Canyon State Park and Historic Site. In her autobiography, *Jane's Window: My Spirited Life in West Texas and Austin*, Jane reflects on that time: "We soon discovered that it can take years of patient persuasion to get the legislature to create a new park, even one containing works of art found nowhere else in the world. To speed up that laborious prospect, Rosie [Rose Mary Jones] worked on the ranchers, and I worked on the politicians. Solveig provided us with the scientific and archeological backup.

Somehow it all came together. Between 1973 and 1977, the State of Texas bought 2,172.5 acres along Seminole Canyon."[60]

When the park was finally completed and ready to open in February 1980, Rose Mary called the Texas Parks and Wildlife Department to inquire about its plans for a dedication ceremony. Upon finding out that the state did not have money for such an event, Rose Mary and Jane footed the bill and organized a grand event that included Wendell Chino, chief of the Mescalero Apache; Lieutenant Governor Bill Hobby; and an army of journalists.

Jane notes in her book that in 2006, more than forty-one thousand visitors had toured the remote Seminole Canyon State Park and Historic Site.[61] Not only do visitors get to experience the canyon's rock art and learn about the Archaic people who created it, they can also connect with the unique landscape of the region. Its windswept rocks—dotted with pockets of grasses, cacti, and trees—are rimmed by deep canyons that are home to more than two hundred migratory or resident birds, including Zone-tailed Hawks, Scaled Quail, Painted Buntings, and Canyon Wrens.[62]

In 1991 Jane, D. J., and their friends formed the Rock Art Foundation to purchase and protect a property adjacent to the park called the White Shaman Preserve. The Rock Art Foundation also provided funding for several of Solveig Turpin's research projects on the Native American pictographs, petroglyphs, and archeological sites of the Southwest and Mexico.[63]

Always up for an adventure, in 1986, Jane swam in the fiftieth anniversary of the Fort Stockton Water Carnival. That same year, she and D. J. donated the Parade Grounds of Historic Fort Stockton to stimulate restoration of the historic location- including Comanche Springs. The two had founded the Fort Stockton Historical Society in the 1950s. During the 1980s, the Sibley-Potts Foundation provided a matching grant to establish the forty-nine-acre Sibley Nature Center in Midland to provide conservation education and a connection to nature for schoolchildren and adults. The center has recorded more than 180 species of birds at the location.[64]

Born into moderate means, D. J. and Jane Sibley used their brilliant minds and oil-fed fortune to benefit the communities they loved. It is hard to summarize the expanse of influence the Sibleys have had. Every heart lifted by the rising notes of the Austin Symphony. Every visitor who learns about the rough-and-tumble history of Fort Stockton. Every artist who hones their craft at Laguna Gloria. Every mind enchanted by the pictographs and sweeping landscapes of Seminole Canyon. Every child who connects with nature at the Sibley Nature Center are among the millions who have access to these resources because Jane and her husband gave their time, talent, money, and hearts to the state of Texas.

TEXAS GARDEN CLUBS

Author's note: For purposes of historical accuracy in this chapter, I will use names as they appeared in garden club documents and newspapers. It was uncommon to use women's first names in the early twentieth century; instead, names were listed as "Mrs." followed by the husband's name. If the woman was unmarried, she was referred to as "Miss" followed by her last name. On occasion, a first name might be used if the woman was widowed or divorced. This naming method changed in the 1960s. While researching the women in the early garden clubs and federation of women's clubs it was often challenging to discover the woman's first or full name.

In 1929, just before the Great Depression, conservation gained a strong ally when garden clubs across the state formed Texas Garden Clubs, Inc. That same year, nineteen state clubs, including that of Texas, came together to create the National Garden Clubs. From the beginning,

Members of the Mineral Wells Garden Club, circa 1938. Courtesy Texas Association of Women's Clubs Collections, The Women's Collection, Texas Woman's University, Denton, Texas

clubs placed a robust emphasis on conservation. Independent local garden clubs, most of which were part of the Federation of Women's Clubs, had been in existence since the first one formed in 1891 in Georgia. In 1929 however, the time seemed right for the clubs to focus on their specialties of flower and plant cultivation, flower shows, friendship, and conservation.

In *History of Texas Garden Clubs, Inc., 1928–1948*, author Magnolia Greene described the beginning of the Texas Garden Clubs under its first president, Marian Price Scruggs (1869–1961), known as Mrs. Gross R. Scruggs: "The first annual meeting held as scheduled in Austin in April, formal organization was perfected, and the organizing officers were officially elected. An enthusiastic group of Garden Club members attended from all parts of the state. . . . A six-year conservation and beautification program was adopted—with the efforts to be concentrated on Texas highways in cooperation with the Texas State Highway Department. This was to put highways in gala dress for the celebration of the Texas centennial in 1936. The initial planting was to be redbuds between Dallas and Mineral Wells and wildflower seeds were to be scattered along Highway right of ways."[1]

Similarly, the National Garden Clubs, Inc., introduced a campaign touting the Broadway of America, which was to be a national highway that extended through Texas as it made its way from the East Coast to the West Coast. The highway would showcase wildflowers and native plants in each state it crossed.

Under the direction of its second president, Nellie Roscoe Day Trigg (1888–1956), known as Mrs. Henry Trigg, the Texas Garden Clubs dove deeper into its conservation efforts. The club made a direct correlation between public health and a healthy environment. The club organized trash pickups along rivers and creeks, engaged schoolchildren in planting wildflowers and trees on school campuses, continued beautifying highways with native trees and flowers, and encouraged roadway cleanups.

Volunteer improvement days also were organized in socioeconomically disadvantaged communities, which often, but not always, were communities of color. At first, these improvement programs were well received as providing assistance to create cleaner, safer places for children to play and live. Later, they became perceived as attempts by upper- and middle-class white women to impose their ideals on disadvantaged communities.

Nellie Trigg, perpetual director of Texas Garden Clubs, circa 1930s. Photo included in *History of Texas Garden Clubs, Inc., 1928–1948,* published by Texas Garden Clubs

However, in many communities those projects were taken over by local garden clubs comprising women from those communities.

During this time, as long as the women were beautifying their surroundings, their male counterparts seemed happy to support their causes. After all, beautification was seen as the work of women, while forestry, biology, and even conservation were perceived to be, for the most part, the realms of men.

A February 5, 1933, article in the *Fort Worth Star-Telegram* about the dedication of the wildflower and bird sanctuary at Longhorn Cavern State Park includes a photo of four area garden club presidents along with the chair of the State Wildflower Conservation Committee of the Texas Garden Clubs, Mrs. J. L. Mims (State Parks Board Chairman David Colp joined them at the dedication but is not pictured). The women are shown wearing full-length fur and wool coats and holding clusters of redbuds tied in burlap bags. Today, the image seems odd; however, these women were part of the social strata of their respective communities and wore the fashions befitting their time and station. No doubt they would have never considered allowing the newspaper to photograph or print an image of them looking disheveled or dirty from gardening. This further propagated the false ideal that it was acceptable for women to beautify—and look beautiful in nature—but not appear rugged or tough.[2]

In 1933, under Mrs. Trigg's presidency, the Texas Wildflower Law was passed by the Forty-Third Texas Legislature, which made it unlawful for any adult (eighteen years or older), firm, or corporation to injure or destroy trees, shrubs, or flowers located on public properties without obtaining prior permission. Mrs. Trigg went on to serve as director of the South-Central Region and, later, recording secretary of the National Garden Clubs. In all of these roles, she leveraged her position to advocate for healthy communities, conservation, and the beautification of highways through native plants.

Perhaps one of her greatest gifts to Texas came in 1940 when she was appointed to the Big Bend Park Association. The *Corsicana Daily Sun announced* her appointment on August 29, 1940: "Mrs. Trigg is the perpetual director of the Texas Federation of Garden Clubs and is the only person in the United States to hold that title. Mrs. Trigg has been interested in the Big Bend Park project since its inception. . . . Aside from her activities with the garden clubs of the nation, Mrs. Trigg is also a director in the American Civic Planning Board headed by Delano Roosevelt, uncle of the President."[3]

In the 1940s the Big Bend National Park Association was formed to support the establishment of a national park in Texas. In the 1930s the Texas legislature had quickly cobbled together properties in the region to form a state park in order to take advantage of the Civilian Conservation Corps to construct buildings and roads. The intent was eventually to turn it into a national park, which came with better funding and more prestige. However, there was a large contingency of political leaders who were opposed to the federal government owning or operating public lands in the state—a holdover from the Reconstruction Era. Championed by Amon Carter (1879–1955), founder and publisher of the *Fort Worth Star-Telegram*, the association was tasked with raising money as well as political and public support. The funds were to be used to purchase the additional lands.[4]

Mrs. Trigg toured Texas with fellow Big Bend National Park Association member Jeffie Davis Pringle Wardlaw (1878–1968). Mrs. L. J. Wardlaw, a former schoolteacher, was appointed to the association in 1940 after serving for several years on the Big Bend Park Committee of the Texas Federation of Women's Clubs. As publicity chair of the committee, she published a booklet that was sent to all eleven hundred clubs in Texas describing the scope of the national park project and explaining ways they could host fundraisers for the cause. It is also reported that she toured much of the proposed Big Bend National Park on horseback and by hiking. She published the booklet and paid for all of her travel out of

Big Bend Park Program, *Fort Worth Star-Telegram*, February 25, 1940. *Left to right:* Nora Anderson, Jeffie Wardlaw, and Frances Talbot. Courtesy Special Collections, The University of Texas at Arlington Libraries

her own pocket. Her husband, Judge Louis J. Wardlaw, enthusiastically supported the causes and efforts of his adventuresome second wife.[5]

The two powerhouse women arranged to show a color motion picture by Jack Lamb on the splendors of Big Bend. The film was shown at garden clubs, federation clubs, rotary, town halls, and other events throughout 1941. The movie was one of many tools the association used to engage community leaders by showing them what their influence and investment could yield.[6]

In March 1941 at the Texas Garden Clubs' annual convention, Mrs. Trigg and Mrs. Wardlaw made the case to support the legislation needed to create the park and financially give to the movement. On March 13, 1941, the *Brownsville Herald* described Mrs. Wardlaw as "painting a word picture of a huge international park to stand as a symbol of permanent peace between the United States and Mexico." The article quoted Mrs. Wardlaw as saying, "Nature herself seems to have decided upon the Big Bend as a perpetual wilderness. Remaining isolated from the outside world the region

Mrs. Hugh Stewart (*left*), Big Bend Park Committee chair of the Jacksboro Junior Woman's Club, and Mrs. E. E. Farries, club president, with District Attorney John Moore, examining motion picture films of Big Bend National Park. Published in the *Fort Worth Star-Telegram* on March 16, 1944. Courtesy University of Texas at Arlington Fort Worth Star-Telegram Collection

retains to the full its primitive characteristics." She went on to describe the ethereal "ghost mountains" and silent drama of the ever-changing light.[7]

Both Mrs. Trigg and Mrs. Wardlaw, along with Big Bend National Park Association board members and longtime advocates, envisioned Big Bend as a living scientific laboratory to be used to study and understand all aspects of nature and geology. They also hoped it would become a destination for artists, writers, and tourists who would infuse local economies as they traveled to and stayed in the area. The Texas Garden Clubs and Texas Federation of Women's Clubs endorsed the project and threw the weight of their memberships of more than fifty thousand women behind the creation of the first national park in the state. Big Bend National Park opened in 1944 and has held true to that early vision every day since then as people come to enjoy the natural splendors of the region.

Magnolia Greene holding a copy of the *National Gardener*, date unknown. Courtesy The Dolph Briscoe Center for American History, The University of Texas at Austin

Sadly, neither of these women nor the efforts of the clubs are mentioned in history books about Big Bend, including John Jameson's book, *The Story of Big Bend National Park*. Thankfully, a record of their contributions exists in the archives of the garden clubs and local newspapers such as the *Fort Worth Star-Telegram*.

Magnolia Neighbors Greene (1889–1965), known as Mrs. John W. Greene, recorded much of the first twenty years of the garden clubs in her book, *History of Texas Garden Clubs, Inc.* Mrs. Greene did more than just record the history; she championed the mission of the garden clubs, especially in regard to conservation.

Born in Kentucky in 1889 to John and Melvina Neighbors, Magnolia was one of eight children. The Neighborses moved to Texas in the early 1900s and settled in DeWitt County to farm. Magnolia graduated from Southwest Texas State University and proceeded to teach for approximately twenty years in Texas and Oklahoma. She was around forty-five years old when she married businessman and rancher John W. Greene of Georgia. The late-in-life marriage did not yield children for the couple. Like so many other women who resigned from a career once they were married, Mrs. Greene joined the Federation of Women's Clubs and Texas Garden Clubs for friendship, mental stimulation, and a connection with the outdoors.

Mrs. Greene worked her way through the ranks of the Violet Crown Garden Club of Austin and the Texas Garden Clubs. The Violet Crown Garden Club was founded in 1924 by businesswoman and philanthropist Clara Driscoll (1881–1945). (The club is still active one hundred years later.) In 1945, while president of the Violet Crown Garden Club, Mrs. J. W. Greene won the National Garden Clubs' National Conservation Green Ribbon award for the Save the Texas Bluebells project.

In a July 22, 1945, *Austin American-Statesman* article, Nella Mae Dieter reported, "Bluebells and Buffaloes will not share the same fate in Texas if the movement among Austin gardeners continues to make bluebell disciples of gardeners throughout the state. . . . The bluebell used to cover entire fields in the land surrounding Austin. Many old residents of Austin recall that in years past a drive on a summer afternoon could be made in any direction from Austin to see them. Today, a solitary descendant or two stand in an occasional field. Because of its blue beauty and its lasting quality as a cut flower, this native plant has reached a state of near extinction. . . . A committee from the club made a trip to a grower near Houston and returned with enough bluebell seed to fill two shoeboxes."[8]

It helped that author, conservationist, and fellow club member Bertha Dobie or Mrs. J. Frank Dobie, chaired the committee and wrote passionately about the topic in magazines and other publications. Mamie Rather Powell also served on the committee.

Mrs. Greene ascended to the presidency of the Texas Garden Clubs, Inc in 1947 and served in that role until the end of 1949. As her first order of business, she aimed to move the garden clubs into an era of addressing social issues at the community and state levels. In a March 4, 1948, *Valley Morning Star* article, author Pat Ford reported, "'Growing flowers is fine, but growing good citizens is better,' Texas State Garden Clubs president Mrs. John W. Greene of Austin declared on arrival here Wednesday for Thursday's opening of the three-day, 20th annual convention. With approximately 400 delegates already registered from the state's 416 clubs with a membership of 17,460 Texas women, Mrs. Greene pointed out the tremendous force made available for the program. A tall, composed, rather quiet woman, Mrs. Greene indicated the primary objectives of the organization were not being abandoned, but rather expanded to include new types of services."[9]

Mrs. Greene's second order of business was to create the *Lone Star Gardener*, a magazine to promote club activities and gardening. She edited the magazine for many years. Third, she advocated for the creation of the Audubon Nature Camp of Texas, which was conceived in 1946 as a partnership between the Texas Garden Clubs and National Audubon Society. The first camp was held in Kerrville in 1948. The Audubon Nature Camp was a five-session, two-week intensive "natural leaders" training camp where scouts, junior club leaders, and aspiring teachers ages eighteen and older could learn how to incorporate nature in their lessons and become advocates for conservation. Alongside fellow club members Mrs. J. Frank Dobie, Mrs. Jerry Stillwell, and Mrs. George S. McGhee, Mrs. Greene worked to launch and sustain the camp at the Schreiner Institute in Kerrville. The women assembled an all-star team of experts to lead the programs, headed by Fred R. Cagle, professor of zoology at Tulane University.

"These young adults will gain the foundation knowledge necessary for teaching across the state about native animals, fish, birds and plants," declared the *Houston Chronicle* on July 10, 1949. Attendees could earn two credits toward their college degree at any college or university in Texas. Each "adventure"—as they called the lessons—was designed to foster an understanding of nature and show attendees how to educate others about it. They also were trained on how to advocate for conservation once they returned home.[10]

Newspapers from El Paso to Houston ran articles announcing the program and encouraging youth to register for the camp each year. The camp gained national attention and won the Bronze Medal, the highest award a state could win, from the National Garden Clubs. Following the success in Texas, other camps were established in additional states.

Mrs. Greene went on to serve as the third vice president of the National Garden Clubs from 1949 to 1951 and remained involved at the national level as chair of the Conservation Committee.

In 1952 she organized an Audubon club in her hometown of Austin to focus on birds, habitat, and conservation education in Central Texas. The first planning meeting was held in her home in February 1952. The City of

Advertisement for the Audubon Camp of Texas, date unknown. Courtesy The Dolph Briscoe Center for American History, The University of Texas at Austin

Austin Recreation Department, under the direction of Mr. Beverly Sheffield, sponsored the society. Attending that first meeting was a who's-who list of conservationists and civic leaders: Mrs. J. Frank Dobie, Mrs. Roy Bedichek, Mrs. I. W. Popham, Mrs. P. C. Oates, Mrs. J. B. Rutland, Mrs. Vera Digby, Mrs. E. T. Dawson, Mrs. Theresa Carrell, Miss Ruth Allen, Mr. Bradly Davis, and Mr. Everett Dawson.

The new club's big idea was to host screenings of a series of color motion pictures offered by the National Audubon Society as a means to capture the public's interest. As the group officially organized, they elected Mrs. Greene as president, Ralph A. Bickler as vice president, and Miss Dorothy Flury as secretary. The group had lofty and interesting goals, including designating the city of Austin a bird sanctuary, creating a bird checklist for Travis County, and creating a nature museum near Zilker Park. The club also organized nature walks along Bee Creek, which planted the seeds of the initiative to preserve what later became the Wild Basin Preserve. By 1953, the organization had adopted the name Travis Audubon Society. To date it is the longest-running conservation organization in the Austin area. In 2024 alone they offered 199 classes and field trips, led by their army of more than 275 volunteers and it manages three sanctuaries with a total of 737 acres.[11]

The seeds for a garden center or nature museum had been planted in 1946 when garden club member Mrs. W. Bradfield approached the City of Austin about donating one of her properties for a center. It was not until 1955 that the Austin Area Garden Center officially created a master plan to raise money for a world-class municipal horticulture center that would serve as an education center and living classroom.

Using her organizing skills, Mrs. Greene planned spring flower shows from 1955 to 1962 as one method to raise money to build the center. She and a team of volunteers worked with the city and private donors to raise the funds. Her legacy, and the legacy of all those who were involved with creating the Zilker Botanical Garden, lives on at 2220 Barton Springs Road in Austin. Each year, the garden is visited by thousands of people and utilized by the local garden clubs for their meetings and events. (One of my favorite meeting spaces in Austin is the Magnolia Greene Room located in the stone building tucked among the twisting live oaks.)

Opening of the Audubon Sanctuary in Austin, April 25, 1954. Lieutenant Colonel and Mrs. Schwartz assisting Mayor McAden in cutting the ribbon. Magnolia Greene, far left in flat hat. Photo by Bradley Davis. Courtesy Travis Audubon

Mrs. Magnolia Greene's obituary states that she was included in the first edition of *Texas Women of Distinction* and was named in the fourth edition of *Who's Who in American Women*. She was honored with multiple awards, and she became the namesake of the Magnolia Greene Garden Club of San Antonio and Magnolia Greene daylily. In 1965 she was named the Outstanding Garden Club Woman of the Year. Sadly, she passed away before the awards ceremony. She died one month before President Lyndon B. Johnson signed the Highway Beautification Act of 1965. However, her papers at the Briscoe Center for American History at the University of Texas in Austin make it clear that she had been following the progress of the bill.[12]

As president of the Texas Garden Clubs in 1963, Mary Kittel (1904–1993), known as Mrs. Howard S. Kittel, compiled, edited, and published *Stepping Stones for Garden Clubs*. Mrs. Kittel was president of the Fort Worth Garden Clubs before becoming president of the Texas Garden Clubs. The book provided written guidelines for conservation education programs and a step-by-step guide for the officers and committee chairs to understand program procedures. The book also listed and encouraged

Mary Kittel holding *Flowers in Glass*, which featured six of her arrangements, April 6, 1954. Published in the *Fort Worth Star-Telegram*. Courtesy University of Texas in Arlington Libraries, digital collection

visitation to every known nature center, state park, wildlife sanctuary, and national site in Texas. Based on thirty years of projects, the book captured the objectives of the Highway Beautification Committee:

> Encourage clubs to promote civic beautification and cooperate with other organized community groups working along these lines.
>
> Work closely with litter bug chairman and continue to stress "Keep America Beautiful."
>
> Work closely with legislation chairman in an effort to have unsightly billboards removed from our highways.
>
> Work toward removal or fencing of eyesores such as automobile graveyards and junkyards along highways.

Work toward improved approaches to towns and cities.

Work with all interested groups toward more roadside parks.

Promote the use of native plant materials in highway plantings.

Promote better planned landscape design through the services of landscape architects as members of highway boards.

Encourage each garden club member to take an active part in carrying out these objectives through local support in each community.

Keep clubs informed of regulations of Texas Highway Department and work closely with this department in an effort to promote highway beautification.[13]

This list is included to show how forward-thinking the clubs were and how the seeds the clubs planted took root in federal legislation.

Mrs. Kittel's book also instructed members to reduce oil and gas consumption and be mindful of the laws and regulations surrounding the petroleum industry. The clubs, and the rest of America, were starting to see the long-term impacts of the fossil fuel industry on the environment. However, the clubs also acknowledged that the industry was one of the state's economic engines. Mrs. Kittel stated that at the time, Texas was losing fifty thousand acres of natural space every year to bulldozers. With that in mind, the Texas Garden Clubs stressed the need for action at the individual, local, state, and national levels to ensure that future generations had access to natural areas and wildlife. She also published several books under her full name, Mary Bradham Kittel. One of her books, *Our Vast and Varied Vacationlands* advocated for nature tourism.

As president of the Fort Worth Garden Club in the 1950s, Mrs. Kittel was part of the committee to build the new Texas Garden Clubs state clubhouse next to the Fort Worth Botanical Gardens. Completed in 1961, Texas became the first to have a state clubhouse. Mrs. Kittel was world renowned for her flower arrangements and traveled to multiple countries and forty-eight states to study techniques and speak on the topic.

In 1963 she represented the Texas Garden Clubs at the US congressional subcommittee hearings on water pollution control and abatement to advocate for clean water and water conservation initiatives. She gave a detailed report of the Texas Garden Clubs' concerns about discharge

Mrs. Reed Sass, Mrs. Howard S. Kittel, and Mrs. Guy Rogers Jr. study a model of the state clubhouse, November 7, 1958. Published in the *Fort Worth Star-Telegram*. Courtesy University of Texas in Arlington Libraries, digital collection

from wastewater plants, agricultural chemical runoff (which was killing livestock), and the growing need for water conservation. She was one of the only women included in the hearing where she took a grilling from the congressmen but held her own.[14]

In 1964 she attended the Governor's Conference on Natural Resource Management and Development in Dallas. At the conference, she gave a copy of *Stepping Stones* to both Texas Governor John Connally and US Secretary of the Interior Stewart L. Udall. This was significant, as Udall was a key adviser to Lady Bird Johnson.[15] Mrs. Kittel encouraged thousands of members of the Texas Garden Clubs to support the Highway Beautification Act. In addition to meeting several times with Governor Connally, she wrote letters to him requesting he take action to establish more state parks, which he did in 1967 by securing a state bond package to improve and purchase additional state parks.

In 1972 she traveled the country to deliver a program she titled "Citizen Action for a Quality Environment." In 1973 she was elevated to the presidency of the National Council of State Garden Clubs and was chair of that organization's American Land Trust Committee. While in that role, she wanted to change the image of the garden club to make it more relevant. She stated, "Men have always thought of us as little old ladies in tennis shoes putting petunias in the wrong places, or wearing white gloves and sipping tea. Actually, we have been quietly doing what the world is just waking up to, the fact that must be done—practicing wise use of our natural resources."[16] The American Land Trust is also credited for helping to establish the Wild Basin Preserve in Austin.[17]

The power of the women's clubs began to wane as women stepped out of the home and into the workforce in greater numbers during the 1970s and 1980s. Some of the larger national projects that they supported, including the Migratory Bird Act, Clean Water Act, and Clean Air Act, were passed into law and being enforced. As other conservation groups stepped forward in the 1970s and 1980s, the garden clubs shifted their focus to play a supportive role rather than take the lead on large-scale conservation projects. Today, one of their primary objectives is still to secure conservation education in school curriculums and after-school programs. They also continue to plant local pollinator gardens to support native bees as well as migrating butterflies, moths, and hummingbirds.

Women of Color Garden Clubs

While conducting research on the garden clubs, I began to see that the treatment of women of color, African Americans in particular, ebbed and flowed depending on the time and community. The Texas Federation of Women's Clubs and Texas Federation of Colored Women's Clubs had interracial committees in the late 1920s and early 1930s. Those members, as well as the presidents of various community clubs, often attended each other's state conventions and local meetings. The early garden clubs were also members of the respective federations.

The first garden clubs for African Americans often bore the name "Negro Garden Club" of the community in which they were established;

Maids and Matrons Club of Marlin, 1940. Courtesy Texas Woman's University–Denton, Special Collections and University Archives

if they had another name, the newspapers still only referred to them as the "Negro Garden Club." Often, the clubs for women of color were associated with the YWCA or a local church. The Blue Triangle YWCA Garden Club of Houston was founded in 1937 by Mrs. Mollie Parrott (its first president), Mrs. A. E. Bowie, Elouise Williams, Virgie Blake, Viola Burton, Mrs. J. B. Covington, Mrs. R. H. Ward, Carter W. Wesley, Fannie Murray, Tasmania Pemberton, Mrs. J. W. Richardson, and Stella Starks. Now known as the Blue Triangle Garden Club, it is the oldest African American club still operating in the state.

These clubs did exactly the same things that the white women's clubs did—hosting flower shows, holding conventions, sharing gardening tips, engaging in community improvement projects, cultivating leadership skills, and, above all, fostering friendships. Because the newspapers rarely covered the good work being accomplished in African American communities by African Americans, it is difficult to ascertain the extent of those projects. In 1949 there was a mention in the *Fort Worth Star-*

Telegram of the Hillsboro "Negro Garden Club" hosting a campaign to "improve sanitation and beautify homes of colored residents." The article did not elaborate much deeper on the topic.[18]

By the 1950s, race relations were deeply strained at all levels of society—including the garden clubs—as portrayed in a September 1951 article in the *Corpus Christi Caller-Times* that stated, "Negro garden clubs may send observers to the meetings of the Corpus Christi Council of Garden Clubs, members voted yesterday. The Negro Althea Garden Club has requested membership in the city council of garden clubs, but the council decided they could not be admitted because the state organization has refused membership to Negro clubs." The announcement was a sobering reminder of the depth of segregation.[19]

On occasion, wealth or marriage could elevate a Hispanic woman to a level of acceptance in the early women's clubs; however, that same latitude was not given to African American women.

The early bylaws of the Texas Garden Clubs stated that in order to form a new club, the applicant needed the support of at least two other clubs. In many cases, that support was given willingly to clubs for women of color, as was the case for Kilgore, Hillsboro, Fort Worth, and Austin. An April 11, 1954, quarter-page article in the *Austin American-Statesman* by Clare Ogden told of the formation of a new East Austin garden club (the article did not mention the name of the club). The president of the Violet Crown Garden Club, Mrs. T. R. Holbrook, was in attendance to offer support and advice on organizing. The club was part of the East Austin YWCA, and the first meeting was held at the home of Mrs. Alice R. King on East Sixth Street. Charter members included Mrs. C. R. Stewart, president; Mrs. Jerome Hill, vice president; Mrs. Sophia Jackson, secretary; and Mrs. Carroway, treasurer.[20]

While all was not rosy between races and cultures—even in the garden clubs—before and after the Civil Rights Act of 1964, the clubs were a space where women could cross those lines in appreciation of an individual, skill, or technique cultivating a perfect flower or shrub; share seeds; and be supportive of each other's programs. Each point of contact slowly chipped away at fears and stigmas. Like all aspects of race relations in Texas and beyond, it is still a complicated story to tell. Today, all clubs are integrated in terms of gender, race, culture, and creed.

Fort Worth Federation of Women's Clubs members, circa 1953, Texas Association of Women's Clubs Collections. Courtesy the Women's Collection, Texas Woman's University, Denton, Texas

To Beautify and Save

The two terms that the women of the garden clubs used to further their causes were brilliantly selected: *beautification* and *save*. Both words were palatable to their male counterparts and to women who were not involved with the clubs. Men expected women to make the home, the community, and even themselves beautiful. A woman's attendance at a garden club meeting to talk about beautifying the community was unthreatening to the husband-wife or father-daughter power dynamic. However, if women had said they were going to a meeting to rally against billboards along highways, thus blocking a commercial endeavor in the community, the men might not have been as amenable. Or if they had said they were going to petition the government to protect a threatened species, that also likely would not have gone over well. However, if women were "saving" something for future generations, then the concept was believed by men to be right as rain.

The women in garden clubs across the country also understood all too well that elected officials wanted their votes. They knew that the garden clubs and federations of women's clubs wielded incredible power when it came to nominating a candidate or advocating for a certain topic such as the creation of Texas state parks. Women understood the power dynamic and were not afraid to approach elected officials, who were often the husbands or relations of fellow club members.

The safety of the clubs empowered women to practice being political without actually participating in politics. Women championed causes, pushed for their ideas, built alliances, collaborated with community leaders, and strengthened their self-confidence. They shared these abilities with their daughters and granddaughters, who then marched boldly into the 1960s and beyond to seek even greater equality. By the 1960s, even the power of the word *beautify* was shifting as women sought a more equal platform. During her presidency of the Texas Garden Clubs and National Garden Clubs, Mrs. Kittel fought to make the words *conservation* and *beautification* interchangeable.

In 1964 President Lyndon Johnson and Lady Bird Johnson understood the value of the word and concept of beauty when they established the Task Force on Natural Beauty. The task force helped pull together ideas from around the country on how best to "beautify" or, better yet, protect the environment and conserve nature for future generations. Much of Lady Bird Johnson's national agenda on conservation came straight from the foundations of the Texas Garden Clubs and National Garden Clubs, even if she was unaware of all of their programs and years of effort.

Lady Bird often bristled at the word *beautification*, as she thought it was too feminine. As stated in Lewis L. Gould's book *Lady Bird Johnson: Our Environmental First Lady*, "There was always an undeserved tone of apology and supplication about what Lady Bird Johnson did for the environment; it arose from the label beautification. In this way, her campaign became an instructive lesson in how constricted the role was that women, even the first lady, played in American public life in the 1960s."[21]

However, Lady Bird's advisers and the task force agreed that if junk cars and pollution were ugly and unsafe, then they needed the opposite qualifiers of *beautiful* and *safe* to make the concepts easy to understand and palatable to the public and Congress. Lady Bird and the Task Force

First Lady Lady Bird Johnson at the White House Conference on Natural Beauty, May 24, 1965. Photo by Warren K. Leffler. Courtesy US News and World Report Magazine photograph collection, Library of Congress, Washington, DC

on Natural Beauty were about to take the steps and actions that the conservation-minded women of the garden clubs had wanted to take since the 1930s to obtain the golden goose of policy that the Johnsons made possible. Later in life, Lady Bird revealed, "I'll never forgive Lyndon's boys for turning my environmental agenda into a beautification project. But I went ahead and talked about wildflowers so as not to scare anybody, because I knew if the people came to love wildflowers, they'd eventually have to care about the land that grew 'em." She went on to share, "Beautiful language isn't enough. You have to be very smart about what you are doing when talking about the environment."[22]

Today the Texas Garden Clubs continue to promote the love of gardening, floral design, horticulture, civic responsibilities, landscape, and protecting the environment. They achieve this through educational programs for adults and in schools, environmental camp for members, landscape design school, flower shows, and countless civic engagement opportunities. The Fall 2022 issue of the *Lone Star Gardener* featured an impressive number of nature trails, healing gardens, and pollinator garden projects that had been established around the state thanks to the clubs.[23]

A RARE TEXAS WILDFLOWER

Ellen Dorothy Schulz Quillin (1887–1970)

Many wildflowers in Texas have to be innovative to survive the thin soils, extreme heat, and temperamental rains. Like the flowers, Ellen Quillin also had to be innovative to survive and succeed in the male-dominated provinces of botany, business, and scientific education. That tough innovation also served her well as the central figure in the creation of the Witte Museum in San Antonio.

In 1887 Ellen Dorothy Schulz was born in Saginaw County, Michigan, to a farming family of slim to modest means. In *The Story of the Witte Memorial Museum, 1922–1960*, which she coauthored with Bess Carroll Woolford, Ellen described an incident that occurred when she was a young girl pulling carrots from the earth on her family's farm. After a pack of snooty, wealthy girls snubbed her as they passed by in their parents' fancy horse-drawn carriage, Ellen vowed she would get an education and become a teacher. In her community, teachers were revered, respected, self-supporting, and self-sufficient—not dependent on the land.[24]

Many times, Ellen was the first—or only—female enrolled in her college courses. Money was a constant challenge, as her family could not afford her tuition. Most science majors would have had an opportunity to work on campus or in a more scientific capacity; however, those jobs were not open to female students. In these thin financial soils, Ellen adapted and sold boxes of cosmetics for $3.20 to students and women in town. It would not be the last time she used her innovative mind to achieve her goals.

After her stint selling cosmetics, Ellen worked as a teacher in Texas or Michigan during the school year and attended classes during the summer months, with occasional gaps to return home to help on the family farm. In 1918, at the age of thirty-one, she graduated from the University of Michigan with a master's degree in science and subsequently pursued her doctorate at the University of Texas. By 1922, she was only a few hours shy of earning a doctorate in science.[25]

Ellen Quillin in the field with her camera documenting wildflowers and cacti in Texas, date unknown. Courtesy Witte Museum, San Antonio, Texas

She landed a teaching job in San Antonio and, from 1916 to 1923, Ellen served as the head of the science department at Main Avenue High School in San Antonio. In 1923 she moved up the education ladder to achieve the respected title of director of nature study and science for the San Antonio public school system. She also provided summer instruction in systematic botany at the University of Texas. While teaching at the San Antonio schools, Ellen spent her evenings and weekends collecting flower specimens and began writing her first book, *Texas Wild Flowers: A Popular Account of the Common Wild Flowers of Texas*, which was published in 1928.[26] Two years later, she coauthored a book with Robert Runyon, *Texas Cacti: A Popular and Scientific Account of the Cacti Native of Texas*, published by the Texas Academy of Science in 1930. In addition to her meticulous collection of species samples, she also documented many of the flowers through photographs that she took while in the field.[27]

The year 1922 was an exciting and life-changing one for Ellen. First, she met and became acquainted with Roy Quillin, a kindred nature lover. Second, Roy put Ellen in touch with famed naturalist Henry Philemon Attwater. For decades, Attwater had studied and collected species of birds, plants, and mammals from around the world, but mostly from

Texas. His impressive collection had been displayed at world and state fairs, with portions loaned to the Smithsonian and other institutions. Attwater had expressed interest in selling the collection for $10,000. However, after meeting with the charming schoolteacher, he offered it to Ellen for $5,000.

Ellen engaged her friend at Main Avenue High School, Vice Principal Emma Gutzeit, in her ambition to acquire the collection. Emma jumped on board and devised a structured payment plan that made the impossible $5,000 price tag possible. Joined by other teachers and parent volunteers, the two women sent students onto the streets of San Antonio to ask people to pitch in a dime or whatever amount they could afford. By the end of the first event, they had raised more than $800 for the down payment on the collection. Under Ellen's leadership, the nimble group sold bluebonnet seeds, hosted a theatrical play, and held a bake sale—after all, no education-oriented fundraiser would be complete without a bake sale.

Upon securing the collection from Attwater, they housed it in a closed section of the high school. On October 8, 1923, the public was invited to view the beautifully displayed collection. On opening day, more than three hundred people arrived to admire the taxidermy birds, boxes of butterflies, and the largest collection of pecans in the world. Ellen and Emma recognized the potential for a museum; however, the school had limited space. In 1925 Emma arranged for Ellen to meet San Antonio Mayor John W. Tobin to discuss the possibility of the city committing funds and space for the purpose of a museum. Mayor Tobin agreed to donate a parcel of city land and, later, dedicated city funds with the statement, "Tell you what. I'll give you girls $7,500 if you promise never to come back!"[28]

Ellen did, in fact, return to the mayor's office to see him regarding the contract. That's when events took a strange turn. Mayor Tobin said to Ellen, "Young lady, what if you get married? You see, this building will be built for you because you believe in it, and you've sold me on it. But I've got to have someone to run it. There's no one else who knows what it's all about." In those days, it was not socially acceptable for married women to work for pay outside of a family-run business or the teaching profession. While many married women did work, in certain circles, it was still

Students collecting bluebonnet seeds to sell, circa 1922. Courtesy Witte Museum, San Antonio, Texas

not as welcome of an endeavor, as they were expected to care for their husbands and children. Today, it would seem preposterous for an elected official to insist a woman stay single so she could devote herself to a project or profession. Ellen did not balk at the statement. However, it posed a quandary because she was deeply in love with Roy Quillin. She told the mayor she would postpone marriage and make the museum her priority while also balancing her teaching and writing. Ellen seemed to flourish in the heat of multiple projects.

Ellen and the newly formed Museum Association Board boldly moved forward to construct a building for $75,000 in partnership with the City of San Antonio and San Antonio Art League. Then, in an instant, their luck changed for the better. Like a temperamental Texas thunderstorm, money rained down on the project. Upon his passing in 1925, Alfred G. Witte, a wealthy San Antonio businessman and friend of Mayor Tobin, bequeathed the Museum Association Board $75,000 to build and maintain the museum. His will stipulated that the museum be named for his parents, be located in Brackenridge Park, and be used to display works of art, science, and natural history. In addition, he left his extensive coin and currency collection to the museum.[29]

the Witte Museum or the White Shaman Preserve—or by simply admiring the wildflowers and cacti in the Lone Star State.

From 1926 to 1960, Ellen cared for every aspect of the museum. During those decades, she also found time to teach and lecture on the topics of botany and museums. From the museum's inception, she desired to create a space that would be engaging and educational for children as well as adults. Ellen wanted the space to be a living museum, full of activity and enchantment. She also understood the power of art to tell the stories of nature and encouraged nature-focused artists to display their works in the upstairs galleries of the San Antonio Art Museum. Upon her retirement on October 30, 1960, the City of San Antonio gave her a warm send-off with "Ellen Quillin Day."[31]

Ellen also helped found the Texas Academy of Science, Texas Nature Federation, Cactus and Succulent Society of America, and was a member of the San Antonio Women's Club in addition to other clubs and boards. She and her husband never had children, partly on account of marrying so late. However, the two shared a love of nature and often accompanied each other on business or research trips that involved exploring the outdoors or historic sites. Ellen died of a heart attack on May 6, 1970.[32]

Ellen's primary legacy stands at the corner of Broadway and Tuleta Drive in San Antonio and is still open on Sundays—as Ellen would have wanted. Each year, thousands of schoolchildren, local families, and visitors from distant countries pour through the doors to learn about the natural and cultural history of Texas. Ellen touched the lives of the countless schoolchildren she personally taught and in whom she instilled an awe and wonder of nature. The museum continues to be a space where men and women can combine their love of nature, history, and education in a rewarding career. Women such as Hellen Holdsworth, Director of Stewardship, honor Ellen's legacy as they help the Witte Museum expand the many unique ways it brings nature and history alive within the ten-acre campus set along the San Antonio River. Prior to working for the museum, Helen ran the Texas Brigades program with Texas Wildlife Association.

It was not, however, just children whom she inspired. Connie Hagar and the Nature Club in Corsicana referred to *Texas Wild Flowers* as their botanical bible. Later, the Hagars and Quillins became friends and often explored the wilds of Texas and New Mexico together. On more than one occasion, Connie invited Ellen and Roy to the coast to study a rare flower or witness a unique bird.

Roy Quillin also left a legacy. He and Ellen collected an extensive egg and bird collection over several decades before laws were passed that prohibited the taking of eggs or skins for personal collections. In 1973 Roy entrusted his collection of more than ten thousand eggs to the Rob and Bessie Welder Wildlife Foundation in Sinton, Texas, where they are housed in the Quillin Memorial Wing. The egg collection is significant due to its size and because it provides scientists with an opportunity to study the effects of pesticides on egg development. Many of the eggs were collected before the introduction of chemicals such as DDT, glyphosate (used in Roundup weed control), and other harmful toxins.[33]

Like a rare and resilient Texas wildflower, Ellen persisted through proverbial floods and droughts. She adapted, innovated, and never lost sight of her ultimate goal: to inspire others to notice and revere the nature that surrounds them every day.

OUR ENVIRONMENTAL FIRST LADY

Lady Bird Johnson (1912–2007)

You simply cannot talk about conservation in Texas—or the United States, for that matter—without discussing former First Lady Lady Bird Johnson. At least twice a day, her presence surrounds those who commute to work on a state or federal highway.

Few women in our nation's history have had a greater platform from which to champion environmental actions than Lady Bird Johnson. She effectively channeled her role as First Lady into a force for conservation. She was instrumental in passing several groundbreaking environmental laws during her time in the White House, including the Highway Beautification Act of 1965, which was intended to encourage natural enhancements along roadways, limit outdoor advertising and billboards, and clean up junkyards. The legislation, known as "Lady Bird's Bill," was a culmination of decades of effort from the garden clubs, which focused on highway beautification at the local, state, and national levels. Many books have been written about Lady Bird, so this chapter will focus on only a few Texas-specific highlights of her remarkable achievements. Additionally, her influence spills over into other chapters in this book.

Thomas Jefferson Taylor and Minnie Pattillo Taylor welcomed Claudia Alta Taylor (Lady Bird) into the world on December 22, 1912. Her father owned a general store in the small East Texas town of Karnack. Claudia lost her mother at the age of five and was cared for by a nursemaid, who is reported to have given her the name Lady Bird.

Lady Bird graduated from the University of Texas with a degree in history and journalism in 1934, the same year she met and married Lyndon Johnson. Reports state that Lyndon was taken with Lady Bird the moment they met and was determined to marry her. Indeed, they were married only a few months after their first date in Austin.

Following her father's entrepreneurial spirit, Lady Bird used the inheritance she received from her mother to buy a small radio station in 1943. Over the decades, she would acquire several more radio stations in

Lyndon and Lady Bird Johnson in wildflowers at LBJ Ranch near Stonewall, Texas, 1968. Photo by Frank Wolfe. Public domain, White House Photo Office Collection, Lyndon Baines Johnson Presidential Library, Austin, Texas

Austin, Corpus Christi, and Waco. She remained active in the family business until the last few years of her life.

While her interests and passions ranged far, it was her love of nature that kept her grounded and centered during the highs and lows of public life. According to the Lady Bird Johnson Wildflower Center's website, "Lady Bird Johnson worked tirelessly for conservation and the environment. More than two hundred laws related to the environment were passed during the Johnson Administration, many of which are credited to Mrs. Johnson's work. Among the major legislative initiatives were the Wilderness Act of 1964, the Land and Water Conservation Fund, the Wild and Scenic Rivers Program, the 1965 Highway Beautification Act, and many additions to the National Park system." The website goes on to report, "For fifty of those major initiatives related to conservation and beautification, President Johnson thanked his wife on July 26, 1968, for her dedication by presenting her with fifty pens used to sign these laws. She also received a plaque that read: 'To Lady Bird, who has inspired me and millions of Americans to try to preserve our land and beautify our nation. With love from Lyndon.'"[34]

President Lyndon B. Johnson handing Lady Bird Johnson a signing pen at the signing ceremony for the Highway Beautification Act, October 10, 1964. Public domain. Courtesy Library of Congress, LBJ White House Photo Office Collection

During the years that Lyndon was a congressman for Texas, the couple spent a lot of time on the road driving back and forth between Texas and Washington, DC. Lady Bird noted the rapid changes along the highways as more junkyards and billboards began to diminish the scenic vistas. Twenty years later, she would be in a position to tackle that issue.

The Highway Beautification Act of 1965 was significant for many reasons. It pitted the large and thriving businesses of outdoor advertising and junkyards against those who valued scenic and natural vistas, limited light pollution, and clean water. Junkyards often contained oil and chemicals that leached into the water table or creeks.

The bill did not move through Congress easily. President Johnson told his staff, "You know I love that woman and she wants the Highway Beautification Act and by God we're going to get it for her."[35] As Lady Bird waded into the legislative process, she began shifting the traditional role of First Ladies and testing the boundaries. Her entrance into the political arena was quickly attacked by men such as Congressman Robert Dole, who bristled at the notion of a president and First Lady working as a team and equally capable of generating solutions for the myriad issues facing the country they served. In Montana, a group went so far as to install a billboard that said, "Impeach Lady Bird." [36] She was not only testing the boundaries of the role of the First Lady, she was also testing male-dominated American values. The Highway Beautification Act was signed into law with great fanfare on October 22, 1965. The act remained under constant attack from the day it became law.

President Johnson did not seek a second term and the Johnsons departed Washington, DC, at the end of his presidency in January 1969. Back in Texas, Lady Bird and a group of friends sponsored a collection of awards in the categories of use of wildflowers, enhanced roadside parks or vistas, and maintenance of roadside improvements. The awards were designed to engage the Texas Highway Department to share best practices through friendly competition. That legacy continues today as more native trees and plants are being planted along highways to reduce heat island effects, sequester carbon, and reduce the need for constant mowing.

Lady Bird joined forces with her Austin based friends Ann and Roy Butler, Roberta Crenshaw, and others to shape the parklands and trails around the stretch of the Colorado River in Austin known as Town Lake. She and her team thought Austin, the capital city, should celebrate the land along the river as a space where people could be active, enjoy an outdoor concert, and connect with nature. By 1971 the river's shores had become littered with trash and were thick with brush, and the waters were unusable for recreation because of pollution and debris from the construction of various bridges. After five intense years of planning and raising money, on July 4, 1976, the committed group dedicated a lakeside hike-and-bike trail, which has been a cornerstone of Austin's magnetism ever since. Lady Bird once said, "Each time I walk along the shores of the river, I come away feeling deeply satisfied in seeing our dreams becoming a reality, and Austin a more pleasant place to live and visit." It was suggested many times to change the name of Town Lake to Lady Bird Lake. She always declined the honor; however, just a month after her passing in 2007, the City of Austin passed a resolution to change the name.[37]

In 1965 a collection of friends raised the funds to purchase 269 acres across the Pedernales River from the LBJ Ranch. The land was donated to the Texas Parks and Wildlife Department and opened as Lyndon B. Johnson State Park and Historic Site in 1970. The park offers visitors a chance to learn about what life was like for the early families, such as the Lyndon Johnson's parents, that settled the rugged Hill Country.[38]

On January 22, 1973, President Johnson passed away from a major heart attack. In 1972, before his death, the family donated the "Texas White House"—the nickname for the family's ranch in Stonewall—to the National Park Service. The donation retained a life-estate provision for Lady Bird to live part-time in the family home until her death in 2007.

In 1976 a granite dome known as Enchanted Rock, which is located northwest of the LBJ Ranch and the town of Fredericksburg, was listed for sale. The potential sale of the property captured the attention of several conservationists, including Lady Bird.

The large granite dome is thought to have earned its name from the Lipan Apache, Tonkawa, and Comanche who inhabited the area as they followed the migrating herds and changes in season. The rock makes a variety of noises as it heats, cools, expands, and contracts in the ever-changing Texas weather. The sounds led the first peoples and, later, the Spanish explorers to consider the rock to be enchanted by spirits.

On a hot day in August 1873, the last battle between Anglo settlers and Native Americans occurred in Texas at Packsaddle Mountain. Eight settlers, led by James Moss, a local rancher, defeated a group of Lipan Apache who were hunting along the South Llano River and Beaver Creek. With the Native Americans removed from the area, James Moss bought the land surrounding Enchanted Rock and started a cattle operation.[39]

In 1927 the Moss family opened the land around the rock to the public as a private park, which it remained for almost fifty years. In 1976

Charles H. Moss and his wife, Ruth, had too many health issues to continue managing their ranch and private park operations. The couple hoped to sell the land to a conservation buyer to manage it as a park. However, a conservation buyer did not appear quickly, so in 1977 Charles indicated that he and his wife would sell to any buyer. The Nature Conservancy was contacted but also initially passed on the project, remarking that the organization was not interested in running a park or managing a concessionaire. Two problem solvers stepped up to the plate: James Bryce, a local Austin attorney connected with The Nature Conservancy, and Lady Bird Johnson.[40]

The Moss family had contacted the Texas Parks and Wildlife Department as early as 1962 to indicate they would be interested in selling the significant geological domes and surrounding property to the state. As the department inventoried the state in the 1960s to consider lands they wished to set aside as parks or state natural areas, Enchanted Rock ranked a mere fifty-third on the list. On the other hand, Lady Bird felt passionate about the Hill Country gem. She knew how to leverage her connections and, most importantly, understood the funding mechanisms of the Land and Water Conservation Fund that she had helped enact.

The first lever she pulled was a phone call to Patrick Noonan, president of The Nature Conservancy. When Lady Bird Johnson calls and says, "I need you to come to Texas," you get on a plane and go, which is exactly what Noonan did. She explained the urgency of the situation: the failing health of the landowners, the lack of interest from the Texas Parks and Wildlife Department, and the potential of a mining and quarry company purchasing the property. Noonan saw the natural and cultural significance of Enchanted Rock and vowed to take action.

Lady Bird worked in the background with her influential friends to garner the interest of the Texas Parks and Wildlife Department and move the project up the list. There was always tension between public land proponents and those who thought the state should not own or manage public lands despite the state becoming majority urban in the 1950s.

At the urging of Lady Bird and Pat Noonan, the US Department of the Interior requested that matching funds from the Land and Water Conservation Fund be reserved if The Nature Conservancy could secure the purchase, with the understanding that the Texas Parks and Wildlife

Department would take ownership of the property. With all the pieces in place, the 1,640 acres were acquired on March 1, 1978, with James Bryce managing the transaction. On March 6, 1978, the Texas Parks and Wildlife Department purchased the property from The Nature Conservancy, positioning it to become one of the first state natural areas in Texas.[41]

Enchanted Rock State Natural Area is one of the most popular parks in the state park system. It enjoys the titles and protections afforded it by its National Natural Landmark status and listing on the National Register of Historic Places. In 2025, using money from the voter-approved Sporting Good Tax and Centennial Fund, the Texas Parks and Wildlife Department purchased two additional properties of more than 3,650 acres adjacent the park. Without Lady Bird, no Texan today would have the privilege of climbing to the top of the granite dome to gaze out over the rugged Texas Hill Country and see for a moment what the Tonkawa, Apache, and Comanche saw in their day.

Lady Bird embarked on a new project in 1982. She gave sixty acres in eastern Travis County near Austin and $125,000 to start the National Wildflower Research Center. She recruited her friend, actress Helen Hayes, to cochair the project. The two invited their friends, including the Rockefellers, to donate funds. The research center would study wildflowers, with a particular focus on examining how native flowers and grasses reduce erosion along the slopes of roadways and limit the need for excessive mowing. The highway department was in favor of the research and readily assisted. US Senator Lloyd Bentsen of Texas secured federal funds to encourage the use of native plants along federal highways.[42]

By the 1990s, the center was outgrowing its East Austin space. Lady Bird again stepped in and purchased a new space in Southwest Austin. Today, the 284 acres offer a space for people to connect with nature, conduct valuable native plant research, attend trainings and lectures, or participate in events. The on-site library and website offer a repository of information for those wishing to study various plants and ecosystems in order to better shape our world.[43]

Lady Bird Johnson's legacy touches almost every aspect of conservation in Texas and the United States. She changed the role of First Ladies so that they can pursue legislation and projects that are meaningful to them. To her, social justice and protection of the environment were one

Lynda Johnson Robb and Luci Baines Johnson with Lady Bird Johnson (*left*) at the dedication of the Lady Bird Johnson Wildflower Center, December 22, 1982. Photo by Frank Wolfe. White House Photo Office Collection, Lyndon Baines Johnson Presidential Library, Austin, Texas

and the same. In her mind, no great society could flourish without being moored in clean water, air, and lands.

Several wonderful books have been written about Lady Bird Johnson and delve deeper into how she achieved her many triumphs. To honor her legacy, take the time to visit and explore Enchanted Rock State Natural Area in March and April when the flowers along the highways are at their most brilliant or visit the Lady Bird Johnson Wildflower Center, LBJ Ranch National Park, or LBJ State Park and Historic Site in Stonewall. The Johnsons also left a legacy in the Lyndon B. Johnson School of Public Affairs at the University of Texas at Austin, which has made major contributions to conservation policies in Texas.

advisory board of the Caesar Kleberg Wildlife Research Institute at Texas A&M University–Kingsville, they began to fall in love with the ranch's hardscrabble landscape. Driven by a desire to heal the land and reestablish native plants, Ellen and Buddy used their ranch as a place to conduct research and learn about the wildlife and plant communities of South Texas.

Despite the painstaking removal of mesquite and overabundant prickly pear cactus, the native grasses were slow to return, and some species were not returning at all. Ellen and Buddy saw a need for certified native seeds grown on a commercial scale. Working with the Kleberg Institute, Ellen became one of the founding members of South Texas Natives, with Paula Maywald as the first director. The Temples opened up their property as a living laboratory, with six miles of pipeline scar dedicated as Ellen's Pipeline Prairie. As the project grew and enough seeds became available, it was time to approach the Texas Department of Transportation. Ellen credits state senator and former Texas Department of Transportation Commission Chair Robert Nichols with shifting the policies of the agency to use native seeds along highway rights-of-way.[52] The result of a twenty-five-year effort, thirty-five native species are now included in South Texas Natives' mix of flowering prairie and grasses.

With the assistance of ranch managers Robert and Jenny Sanders, the Temple Ranch developed native habitat to support Wild Turkey, Northern Bobwhite, and many other native South Texas species that had become scarce in the overgrazed landscape. With support from Robert and Jenny, the Temples hosted a program called Notice Nature. The program provided every seventh grader in Duval County the opportunity to come to the ranch to learn about wildlife, soil and water management, plants, hunting, ranching, and wildfire. Game wardens and experts from Texas A&M University, the Texas Forest Service, the Texas Wildlife Association, and other organizations instructed the children and talked to them about careers in conservation. In recognition of their collective efforts on the ranch, the Temples won the Texas Parks and Wildlife Department's Lone Star Steward Award in 2007 and Texas Leopold Conservation Award in 2011.

Ellen joined the Philosophical Society of Texas in 1990 with friend Pat Hayes. In 2001 she was elected president and was placed in charge of developing the theme and assembling speakers for the organization's 164th annual symposium. Ellen's daughter, Susie, describes that event as "a pivotal moment—a window into the evolution of her thinking about the land and conservation."[53]

The theme for the symposium was "The Land," and Ellen assembled a who's-who list of respected speakers including climatologists, soil and water conservationists, policy experts, and writers. She wanted them to look at the challenges the land faced as the world stepped into the twenty-first century. Her goal was to challenge attendees to consider those issues not just in terms of causes and actions but through the deeper lens of the heart.[54]

The year 2001 was globally significant in that the world's population had doubled in thirty years (ahead of projections) to more than 6 billion people. In that same thirty-year period, Texas' population grew from 11.5 million to 21.3 million. Ellen reflected, "Land fragmentation, loss of habitat, draining of wetlands, loss of species—it was all happening at an alarming rate."[55] Following the symposium, Ellen decided that conservation of land, water, and wildlife was the most important cause she could champion. It changed how she saw her role as steward, advocate, and influencer. She and Buddy doubled down on conservation efforts across their private lands and in their philanthropic realms.

In 2006 Arthur Temple Jr. passed away. Cooperate raider Carl Icahn was already making overtures to pick apart Temple-Inland and sell off the assets. The board decided to take a proactive stance and sold more than 1.6 million acres of carefully managed timber forests in East Texas. The sell-off was controversial, but it was on Temple-Inland's terms.

In 2013, as chairman Buddy encouraged the T.L.L. Temple Foundation trustees to buy back nineteen thousand acres that had been sold to the International Paper Company. With the intent to conserve it for future generations, the land was placed into a conservation easement with The Conservation Fund and later moved to Texas Land Conservancy. The Boggy Slough Conservation Area became a new focus for both Ellen and Buddy to encourage innovative and best practices in forest and land

Ellen and Buddy Temple at the tallest and largest loblolly pine in Texas, Boggy Slough, 2013. Courtesy Ellen Temple

management.[56] With directors Steve Jack and Robert Sanders applying innovations and building on years of careful management, Boggy Slough Conservation Area won the Texas Leopold Conservation Award in 2025.

The Boggy Slough property encompasses eighteen miles of the Neches River and its tributaries, which reconnected the Temples to the multitude of challenges facing the river. The fight to save the Neches River started in earnest in 1985, when the Texas Conservation Alliance began looking at mechanisms to preserve the free-flowing waters of the 416-mile river.

A resurgence of interest occurred in 2006 after Richard Donovan published his book, *Paddling the Wild Neches*. Ellen paddled the river several times with Richard, both during the writing of his book and after. Ellen was already connected with Janice Bezanson, Michael Banks, Gina Donovan, and additional champions of the river when she decided to become part of the coalition working to designate the Neches a Wild and Scenic River. The federal designation would prevent selected section of the river from being impounded for reservoirs, among other restrictions.

In 2012 US Senator Kay Bailey Hutchison introduced a bill calling for a formal study to be conducted, which was the first step in earning the wild

Ellen and Buddy Temple (*center*), surrounded by their children and grandchildren at Boggy Slough, fall 2014. Courtesy The History Center, Diboll, Texas

and scenic designation. Unfortunately, the bill did not make it out of the Senate. The Texas legislature also made it clear that the state would not allow additional wild and scenic designations beyond the existing one on the Rio Grande. Ellen and Buddy did not lose hope but saw an opportunity to conserve a portion of the river and the East Texas ecosystem under the auspices of the newly formed Neches River Wildlife Refuge. The T.T.L. Temple Foundation gave $2.5 million toward the acquisition of lands for the refuge. The acquisition was spearheaded by Julie Shackelford at The Conservation Fund.[57] Ellen also worked with Julie to create the Green Infrastructure Plan for Angelina County.[58]

In 2013, within a month of each other, both Buddy and Ellen were diagnosed with cancer. During the fight for their lives, Buddy embraced the effort to reshape Boggy Slough Conservation Area into a place for environmental study, much like the South Texas ranch had become. Sadly, Buddy lost his battle with cancer in April 2015 and died surrounded by the grace and love of his family. With the support of her family, Ellen beat her cancer.

After Buddy passed away and sale of the ranch in South Texas, Ellen reconnected with a project the two had enjoyed together: propagation of native seeds. Honoring Buddy's wishes to have Boggy Slough become a place for research and learning, Ellen reconnected with the Caesar Kleberg Wildlife Research Institute and the Texas Native Seeds Program to establish the Buddy and Ellen Temple Native Plant Endowment in Buddy's honor.

Generating the necessary level of certified seed in various regions takes a commitment from companies to grow and harvest the seed. Through an additional multiyear financial commitment and designated seed plots at Boggy Slough, Ellen led the creation of the East Texas Native Seed Project to benefit the region. The idea was to create certified native seed that could be used by private landowners to restore overgrazed or recently logged areas, by the Texas Department of Transportation to plant along highways, or by contractors to repair pipeline rights-of-way.[59]

One of the benefits of planting native plants along highways is that it reduces the required frequency of mowing, which decreases greenhouse gases emitted by the mowers and eases the mowing budget of the Texas Department of Transportation. Restoring rights-of-way along the thousands of miles of pipeline with native plants instead of nonnative Bermuda grass or KR-bluestem is a game changer in the conservation of wildlife and strengthens biodiversity.

Always learning from the past and looking toward the future, Ellen is considering how landowners can become more informed about the benefits of planting native prairies and how that affects carbon sequestration. "I think what may be missing in the discussion is the fact that a mature prairie is as much of a carbon sink as a forest," she says.[60] Carbon credits and carbon sequestration are emerging topics in Texas but remain on the horizon in many ways. Also on Ellen's horizon is another piece of her past: the Neches River. She and others are considering what it would take to utilize the green infrastructure of the river to create a wildlife corridor. To celebrate the wildlife, wild spaces and wild people of the river, Ellen produced a documentary *The Neches River: Wild Heart of East Texas*, with film maker Curtis Craven in 2025.

Ellen's legacy is almost incalculable. She and her husband, Buddy, have given millions of dollars in financial support to conservation efforts

Ellen Temple paddling the Neches River, 2013. Photo by Jonathan Gerland. Courtesy Jonathan Gerland

around the state, both in private gifts and through the T.T.L. Temple Foundation. In addition, Ellen has given her time and brilliant mind to dozens of organizations and projects that have resulted in valuable research to make the case for native plants, habitat conservation, environmental awareness, and nature education. When and where she could, she has leveraged her political friends to push for conservation-minded policies, the creation of wildlife refuges and nature centers. Additionally, she has been a champion of women's history and women's equality. Her living legacy continues with her children and grandchildren.

In 2015 Ellen was one of the inaugural honorees of the Audubon Texas Women in Conservation Terry Hershey Award. The event fanned the flames to tell these women's stories. I was not ready then, but when I was, Ellen was the first person my mother and I interviewed for this book.

Sunset with birds coming to roost. Blackcat Lake at Boggy Slough Conservation Area, August 2022. Courtesy Jennifer L. Bristol

political center of Austin, the region has always had an independent bent that reflects the spirit of its people. While they might not agree on many things, they often come together to protect the land and the heritage it yields.

THE LIBRARIAN OF THE BIG THICKET

Maxine Johnston

Maxine "Mickey" Johnston was born in 1928 in Gillam, Arkansas. She grew up on a small farm where she loved to escape into the woods to read books and explore. In a 2019 interview, she stated that growing up on a farm connected her with nature as a child, and that love has never faded. A few years after her mother passed away, her family moved to Beaumont, Texas, when Maxine was fourteen, but, in her words, "it was not love at first sight." She did not care for the coastal prairie's flat landscape or endless marshes. But the forests changed that, and the love began to mature.[1]

Before moving to Texas, Maxine's father owned and operated a sawmill and worked as a timber estimator in Arkansas. Sustainable forest management was not yet being implemented or enforced in the 1930s and 1940s. Maxine said the timber industry had one view: "Get out there and cut those things. There were fields of stumps everywhere back then."[2]

The fields of stumps and roads busy with logging trucks filled with timber disturbed Maxine. Her high school teacher and lifelong friend, Alice Cashen, suggested that Maxine write a paper about the forest; the project had a lasting impact. Maxine graduated from Sam Houston University with a bachelor's in science in 1953 and earned a master's in library sciences from the University of Texas at Austin. In 1955 she went to work at Lamar University as its reference librarian and stayed for thirty-three years; she eventually became the director of the library. Sticking with something is one of Maxine's superpowers. Based on the number of awards she won and professional associations in which she served, her librarian career could have been enough for one lifetime. However, Maxine also filled her days with volunteer work from which we still benefit.

While Maxine was mowing the Cashen family plots at the Guedry Cemetery with Alice Cashen, two men drove up and introduced themselves as Lance Rosier and Larry J. Fisher. Larry wandered the cemetery with them and regaled the two women with stories of the lovely flowers that once grew in the cemetery before it became excessively mowed and

Maxine Johnston archiving materials at Lamar University Library, date unknown. Courtesy Mary and Gray Library, Lamar University, Beaumont, Texas

groomed. Maxine recalls Lance remained in the car and asked her a litany of questions such as, "Are you married? Why not? Et cetera." Lance was a longtime advocate for preserving the Big Thicket and recruited Maxine and Alice to the cause.

Lance worked in the Biology Department of Lamar University and saw the potential in Maxine as someone who could contextualize informa-

Logs to lumber, Southern Pine Lumber Company, Diboll, Texas, circa 1940s. Courtesy The History Center, Diboll, Texas

tion and serve as a bridge between biologists, elected officials, and concerned citizens. She also had one very important device in her office and home: a telephone. Not every office and home had telephone service or could afford to install it, especially in rural East Texas. She used the phone to relay messages to Lance, who had earned the nickname Mr. Big Thicket and was famous for his guided trips into the woods. Maxine's phone, office, and excellent memory for names, dates, and details put her in the catbird's seat as the battle for the Big Thicket marched forward.

The East Texas Big Thicket Association dates to 1936 and consisted of R. E. Jackson, Lance Rosier, Bessie Reid (Mrs. Bruce Reid), Ray Gill, P. A. Winkler, Larry Jean Fisher, Donald O. Baird, H. B. Parks, Walter Taylor, Joe Record, Henry Flagg, William Tucker, and M. L. Yount. This early collective explored the idea of preserving large portions of the forest as a game reserve or national forest.[3] The group's early efforts were halted by World War II. The shipyards along the Gulf Coast put a high demand on the timber and oil industry. The housing boom that followed the end of the war placed an even greater strain on the dwindling forest. The association folded not long after the war, but in 1964 Dempsie Henley and Lance Rosier resurrected it as the Big Thicket Association. Their call to action was to place as much of the thicket as possible into a preserve system of state and national parks.

In 1966 US Senator Ralph Yarborough introduced the first Big Thicket National Park bill. The bill failed but was followed by a succession of other bills over the next eight years. As each bill was introduced, it altered the number of acres and total budget. None of the bills set the boundaries of the parcels to be preserved, which remained part of the controversy. By then Maxine was deep in the trenches, and she traveled to Washington, DC, to meet with elected officials and obtain the maps that accompanied each bill. She developed presentation materials using the maps to help the local public understand what the system would look like.

Maxine utilized the Lamar University archives to warehouse historical documents, plant and wildlife studies, folklore, and stories from the citizens who called the Big Thicket home. She used her extensive knowledge and resources to assemble displays and write papers that were easy to understand and relatable to both the citizens who lived in the region and congressional leaders.

She did not demonize industry, criticize scientists, or dismiss citizens. Instead, she listened to all sides and provided them with the information they needed to make informed choices. Her skills as a presenter and connector were recognized by US Senators Ralph Yarborough and Lloyd Bentsen as well as US Congressmen Charlie Wilson and Bob Eckhardt, who called on her frequently to speak to the media or connect them with local landowners. In a 2018 interview, Ellen Temple stated that Maxine "charmed all the players."[4]

The players were not always at ease with each other. The timber companies were at odds with each other as some, such as Temple-Eastex, saw value in setting aside lands to ensure the health and longevity of the forests, while others felt the whole proposal was a land grab that would cost them revenue. Maxine shared that, at one point, a timber company that she did not want to name even tried to turn the local and state garden clubs, which had been supportive, against the project.[5]

Temple-Eastex was owned by Arthur Temple Jr., who was not in favor of a national park but was not opposed to conservation of the forest. The Big Thicket Association and other conservation groups considered him to be the most moderate and most trusted of the timber barons. In our interview, Maxine shared that "the Big Thicket absolutely would not have happened without the support and influence of Arthur Temple."[6]

The oil companies were equally split and equally angry. Then there were the residents who thought the national preserve would raise their taxes and cost them jobs. Nor did they like the idea of the federal government owning property in Texas. Despite the vying interests, Maxine held the story together and kept each player focused on the near-term, midrange, and long-term goals.[7]

In 1973 Maxine became president of the Big Thicket Association after serving several years on the board. She understood that stepping into that role would be a serious time commitment; however, her employer

Sara Loudermilk, Maxine Johnston, and Representative Charlie Wilson, March 1991. Inscribed, "I'm listening Maxine!" by Charlie Wilson. Courtesy Maxine Johnston

encouraged her to accept. The time was now or never, as reports showed that almost fifty acres per day of Texas forests were being lost to logging and development. As passage of the bill and creation of the preserve seemed imminent, Billy Hallmon noted that the National Park Service had not adequately done the fieldwork on the proposed units. Billy, a resident of Dallas and member of the Big Thicket Association, had skills as a map maker. He drove in on the weekends and walked every unit with Maxine. With topographical maps in hand they collected the ecological data needed and set the boundaries. The maps and data were sent to Representative Charlie Wilson, who directed copies be mailed to the timber companies along with a request that they halt clear-cutting activities in the proposed units.

That same year, Congressmen Wilson and Eckhardt introduced the bill to establish the first-ever national preserve, which is different from a national forest or national park. In 1974 President Gerald Ford signed Public Law 94-439 to establish the Big Thicket National Preserve. But the fight was not over.

The Big Thicket is not a continuous parcel of land like Big Bend National Park. Instead, it is a matrix of several large parcels of forest throughout Southeast Texas. Each unit took time to acquire. Maxine continued to bring the community together as these additional parcels were assembled. Watchdogs had to make sure that the timber companies or rogue loggers did not clear-cut the lands before they entered the preserve system. Even after the boundaries of the preserve were established, the question remained: Could the timber companies have access to the national preserve in the same way they could a national forest? Maxine believed that at least a portion should be off-limits to logging. She and Congressman Wilson did not see eye to eye on this issue, but eventually they came to an understanding.[8] Additionally, several parcels that had valuable ecosystems did not make the final cut in 1974 and required more collaboration to set them aside as state lands or through land trusts.

In 1980 Ronald Reagan was elected president, and soon thereafter he appointed James Watt as secretary of the interior. Secretary Watt demanded that he personally approve every single acre that was added to the federal public land holdings. His policy slowed the addition of critical parcels of acreage to be placed under the protection of the preserve, which further strained the relationship between the Big Thicket proponents and timber and oil companies.

By the 1990s, lands were still being added to the system, and Maxine stepped back into the role of president of the Big Thicket Association as US Senator Phil Gramm tried to derail the addition of ten thousand acres in the Village Creek area. Maxine and the association held the line until finally, in 1993, President Bill Clinton signed the Village Creek Addition Act.

In the end, 112,000 acres of some of the most ecologically diverse lands in the United States were set aside despite fierce opposition.[9] Not all of the lands were purchased by the federal government; Maxine credits The Conservation Fund for raising funds to purchase many of the properties, thus creating a true public-private partnership.

Maxine personally took people on walks into the woods; flew to Washington, DC, to testify before Congress; gave countless interviews; and continued to write and collaborate on articles, books, and journals to raise awareness of the unique region. In 2011 she and Pete Gunter (another strong advocate for the Big Thicket) collaborated with Lorraine

Geraldine Watson outside her home on Lake Hyatt near Warren, Texas, date unknown. Courtesy Mary and Gray Library, Lamar University, Beaumont, Texas

Bonney on *The Big Thicket Guidebook: Exploring the Backroads and History of Southeast Texas*. It was one of many books to which she contributed.[10] Maxine noted that author, naturalist, and folklorist Francis Abernethy called it "his reference book" on the Big Thicket.[11]

Geraldine Watson (1925–2012) was advocating for the Big Thicket at the same time and often alongside Maxine. Geraldine was an artist, naturalist, and writer who had a weekly column in the Hardin County *Pine Needle* newspaper where she shared her vast knowledge of plants and wildlife in a relatable manner. She was a member of the Big Thicket Association and a key partner in raising awareness and leading elected officials, journalists from publications such as *Life* and *National Geographic*, and private citizens on walks to discover the awe and delight of the forest. Maxine reflected in our interview that she and Geraldine shared the same goals of protecting the forest, but approached the details and the public with divergent styles.[12]

In 1974 Geraldine testified before Congress on the economic and ecological value of the Big Thicket. After the national preserve was established, she worked for the National Park Service for fifteen years as a ranger. She authored several books about the ecology and the people of the Big Thicket and donated eleven acres for a sanctuary outside Warren, Texas. The Watson Rare Native Plant Preserve is open to the public and is a magical, peaceful place to immerse oneself in the vibrance of the forest.

In 2019 I read James J. Cozine Jr.'s book, *Saving the Big Thicket: From Exploration to Preservation, 1685–2003*, and found it to be filled with pages of good information about the men who were involved with the Big Thicket National Preserve. Granted, at the time, there were no women elected officials in East Texas who were engaged in the Big Thicket fight. However, Maxine Johnston and Geraldine Watson, two of the preserve's essential advocates, were mentioned in only a few short sentences in Cozine's book. Reading that book helped spur my desire to write this one.

In 2008 Peter Gunter gave a presentation to the Native Plant Society of Texas and stated this about his friend Maxine Johnston: "James Cozine has written a fine dissertation on the second Big Thicket agitation, and I have not the slightest desire to disparage it. I do, however, want to stress

Maxine Johnston (*front*) with protesters, 2022. Courtesy Ellen Buchanan

two contributions to the struggle whose importance otherwise might be overlooked. First is the work of Dempsie Henley, the Big Thicket Association's first president." Then he went on to state this about Maxine: "At the home of Alice Cashen and Maxine Johnston, hospitality flowed without limit, and at virtually all hours of the day and night. If the Thicket Museum headquarters provided a constant public forum, the Cashen-Johnston residence provided a retreat from the tensions of public forum and political friction. . . . It made a profound difference, though as

SAVING TEXAS' WILDERNESS AND RIVERS

Janice Bezanson

In our 2019 interview, Janice Bezanson reflected on how her connection with nature gave her a sense of place and later a focus of purpose. Born in 1950 in Brownwood, Texas, Janice grew up spending idyllic days playing outdoors with her brothers. She learned to swim in the Colorado River at her uncle Bill and aunt Fern's ranch in San Saba County and remembers harvesting pecans with her cousin. One summer, she returned to the river to swim and discovered that it was not flowing; irrigation upstream was pulling water out of the river, leaving it in shallow, isolated pools downstream. Even as a child, Janice felt the injustice of pumping upstream.[16]

After graduating from high school, Janice attended the University of Texas for a year, marrying Steve Bezanson soon after. Steve had grown up in East Texas, which is where the young couple moved to be closer to his family. While in Nacogdoches, Janice attended Stephen F. Austin State University and she and Steve welcomed their son, David, into the world.

Janice credits her husband as being one of the main catalysts for her becoming involved with environmental issues. She shared that, while in college in the 1960s, they both were "magazine conservationists," following *Audubon* and *Friends of the Earth*. An article about creating wilderness areas in East Texas caught their attention. The article asked readers to write letters of support, which is what Janice and Steve did. Ned Fritz, who proposed the idea of the wilderness areas, and Beth Johnson with the Texas Committee on Natural Resources reviewed the letters and replied, asking the Bezansons to volunteer. Steve encouraged Janice to get involved. She told Beth she would be happy to help, but only typing, filing, and answering the phone—nothing public.[17]

Ned Fritz had a propensity for seeing talent and recruiting volunteers. Janice laughed as she noted that Ned "liked to throw people into the deep

Janice Bezanson at the Upland Island Wilderness Area, 2009. Courtesy Janice Bezanson

end of the pool before asking if they could swim."[18] Janice not only could swim, she also could navigate the ego-infested waters of East Texas politics. She progressed from being a volunteer in the office to publicly debating the supervisor of the national forests in Texas on regional TV within a year and testifying before Congress in Washington, DC, a year or two later.[19]

Achieving a wilderness designation is important in preserving sections of national forests and other federal lands from logging, roads, mining, and other extractive activities. Contrary to popular belief, national forests are open to logging and the building of roads into the forests to retrieve timber. Both activities are damaging to surrounding ecosystems. Ned Fritz proposed—and succeed in securing—the wilderness designation in East Texas.

From the moment Janice got involved, she began building coalitions and nurturing relationships. Coalition building is sometimes overlooked as a cornerstone of conservation. However, for long-term system changes and large-scale preservation to happen, grassroots community-driven coalitions are key. The most effective conservation occurs when the will of the community is at the heart of the project. As Janice said, "It's important not to count anyone out. You cannot assume who your allies should—or will—be."[20]

On the surface, reservoirs seem like a good idea. Construct a dam on a river, flood the bottomlands, and—poof!—you have a lake for drinking water and recreation. What Janice—and countless others whom she helped organize—saw was what lay beneath the watery surface. Dam building is big business. With 7,173 dams of six feet or more, Texas ranks first in the country in number of dams. California has 1,468.[21] The lucrative contracts to plan and build a dam are coveted by specialized engineering and consulting firms, which employ powerful lobbyists to bend the ears of elected officials.

Ecosystems are not the only thing affected by creating a reservoir. The people living in the path of the proposed lake are displaced from their homes through eminent domain. And it's not just their homes that are submerged; their churches, cemeteries, and businesses face the same demise. In addition, reservoirs cause the loss of thousands of acres of vital agricultural, pasture, and timber lands, plus habitat for terrestrial, aquatic, and avian species.

During the late 1980s, when Janice and her coalition were contesting the Little Cypress Creek Reservoir outside Marshall, Texas, they inundated the surrounding communities with factsheets and ads that explained the project's negative economic impact, how many people would be displaced, and why a new reservoir was not needed when plentiful and readily available water was housed in the existing thirty or more reservoirs in East Texas.

The Little Cypress Creek project had been placed in the state water plan by the Texas Water Development Board in the 1960s, but funding issues and lack of community support slowed its launch. Because bond money had been pledged to the project from within the Little Cypress Utility District, it had to receive approval from the district's voters. Twice, bond packages were defeated, thanks to the factsheets, TV and radio spots, and press releases that Janice and the grassroots movement generated. Janice pointed out that once people understand they have an alternative to the creation of a reservoir and a voice to speak against it, they can become a powerful force.[22]

During a press conference on the bond package, a reporter held up the Longview newspaper and asked Janice, "Have you seen this?" A full-page ad—titled "Who Is Janice Bezanson?"—had been purchased by the Committee for Water for East Texas's Future, a pro-reservoir group, to discredit Janice: "She [Janice] is a lobbyist who lives and works in Austin. . . . The purpose of this ultra-liberal environmental group is to oppose, by any means available to them . . . every new reservoir. . . . They don't care if you run out of water. . . . They don't care about the economic well-being of your future or that of your children or grandchildren. Their philosophy is, if they can delay each project long enough and cause its price to rise sufficiently, then soon, no one can afford to build a lake. Do not be deceived or confused by their lies. . . . We will not be manipulated by environmentalist extremists from Dallas or Austin!"[23]

Janice defused the situation during the press conference by laughing at the ad. But when she got home, she says she cried, shook, paced around—and then fired back. Because she was a volunteer and private citizen, she could pay for her own rebuttal in the paper. She titled it "I Am Janice Bezanson": "I am not a lobbyist, but merely a volunteer conservationist and private citizen concerned about the human, economic, and environmental future of East Texas. I was invited by numerous local residents to help them oppose an unneeded reservoir. Contrary to accusations, the organization that I chair [the Texas Conservation Alliance] has never lied, deceived, or intentionally confused anyone. We have a 20-year reputation of rock-ribbed integrity and straightforward volunteer citizen effort. . . . One wonders why proponents of the reservoir feel the need to stoop to malicious campaign tactics. Are they perhaps afraid that the voters of Texas have realized that Little Cypress Reservoir is an expensive and unnecessary use of taxpayers' money?"[24]

The campaign to defeat the reservoir succeeded. As a result, people were able to remain in their homes, and thousands of acres of bottomlands were saved from a watery grave.

Janice volunteered for and served on the board of the Texas Committee on Natural Resources, now the Texas Conservation Alliance, for eight years before becoming its executive director, a position she held for more than twenty years. Through her leadership and calm, resilient methods of building grassroots movements and coalitions, she has helped defeat proposals for the creation of five reservoirs that would have flooded more than 125,000 acres of bottomlands and reduced flows downstream along those rivers. Her work in supporting community-driven opposition to the Ringgold and Marvin Nichols Reservoirs continues.[25]

Janice also built local support and advocated for the establishment of four national wildlife refuges (Little Sandy, Neches River, Trinity, and Caddo Lake), as well as defending them from proposed reservoirs and other threats. Opposing the creation of a reservoir or establishing a national wildlife refuge can take decades and, in Texas, is often accompanied by a multitude of political battles and lawsuits. Janice and the Texas Conservation Alliance know that winning the good fight means playing the long game to consistently provide affected communities with science-driven data, economic impact studies, and places for them to voice their concerns.

In the case of the Neches River National Wildlife Refuge, the US Fish and Wildlife Service was pursuing an option to set aside a portion of the bottomlands along thirty-eight miles of the Neches River to support local and migratory wildlife. In 1985 the Bottomland Hardwood Preservation plan documented the significance of the ecosystem along the upper Neches River. The idea of creating a refuge along the Neches was pursued around 2002 when it finally gained traction. In 2004 the US Fish and Wildlife Service began the long process of hosting public comment meetings and informational sessions regarding the proposed refuge. Thanks to Richard Donovan's 2006 book, *Paddling the Wild Neches*, the public became more interested in preserving this iconic and wild space that represented the fading heritage of the region.

Janice Bezanson working the Texas Conservation Alliance booth at EarthX Festival on Earth Day, 2011. Courtesy Janice Bezanson

However, the 1984 Texas Water Plan also identified several proposed reservoirs to flood segments of the Neches River, including some of the same bottomland forests the refuge was hoping to conserve. In 2005 the Dallas City Council passed a resolution approving consideration of the Fastrill Reservoir, which would give the city rights to 80 percent of the water impounded. US Fish and Wildlife requested that the proponents of the reservoir come up with an alternative proposal; however, one was never produced. In June 2006 US Fish and Wildlife was ready to move forward with the plan to establish the refuge. One small issue remained: US Fish and Wildlife needed an interest in the land within the proposed boundaries of the refuge for it to become a reality.

Wanting to help, James and Annie Yount gave the US Fish and Wildlife Service a donation of one acre. The donation gave US Fish and Wildlife the land they needed to move forward with the refuge. In November of that same year, the Texas Water Development Board named the proposed Fastrill Reservoir a priority in the state water plan. In January 2007 the City of Dallas and the Texas Water Development Board sued US Fish and Wildlife, the US Department of the Interior, and the Younts. The case went all the way to the Supreme Court, which dismissed it when the US Third District Court and Fifth Circuit Court of Appeals ruled in favor of the defendants and refuge.[26]

As the executive director of the Texas Conservation Alliance, Janice had established a strong relationship with media outlets in East Texas, and they quickly framed the issue as a "David versus Goliath" dispute: East Texas versus Dallas. A twenty-five-thousand-acre wildlife refuge versus a thirty-five-thousand-acre reservoir. Saving a river and forest versus daming and submerging it. A rural way of life versus an urban one.

For many residents, neither option was desired; they simply wanted to live their lives. Thanks to Janice, the grassroots campaign that started with a group of concerned citizens evolved into a highly functioning coalition led by the Texas Conservation Alliance and the Friends of the Neches River. Other organizations such as The Conservation Fund, the Sierra Club, and the Big Thicket Association all participated to secure the refuge.[27] As of 2022 the Neches River National Wildlife Refuge stood at 7,484 acres of its intended 25,000 acres within the acquisition boundary.

It should be noted that during the 2000s, a radical shift in the management of East Texas forests occurred, making it even more important to create wildlife refuges, increase lands held in conservation easements, and expand public land holdings. During that time, profit-driven boards and corporate raiders were trying to influence most of the major timber companies to sell more than three million acres of forestland in East Texas, much of which was eventually sold to timber or real estate investment companies. In 2007 alone, more than one million acres of land belonging to Temple-Inland was divested. The Conservation Fund's Texas Director Julie Shackelford states that "of all the sales, Temple-Inland's announcement was the most devastating blow to East Texas forestlands because of their generally good land management ethic."[28] Years of positive work with each of the involved companies were lost and had to be reestablished with new landowners, including the now syndicate-owned pieces.

Julie and Cliff Shackelford out birding. Courtesy Texas Parks and Wildlife Department, 2016

Julie Shackelford is a champion of the East Texas ecosystems, including the Neches River. As the director of The Conservation Fund, she has raised over $100 million in public and private funds to protect some of Texas' most iconic landscapes. She and her team have conserved over one hundred thousand acres through conservation easements and fee simple acquisitions that contribute to the Big Thicket National Preserve and the Neches River, the Upper Texas coast at the McFaddin and Anahuac National Wildlife Refuges, the Rio Grande Valley at the Laguna Atascosa National Wildlife Refuge, and state lands including Village Creek State Park, Powderhorn Ranch, and the J. D. Murphree Wildlife Management Area.[29]

In December 2023, under Julie's leadership, The Conservation Fund, with support from the US Fish and Wildlife Service, announced the addition of over three thousand acres to the Neches River National Wildlife Refuge. The funds came from private investment from the federal Land and Water Conservation Fund and the Migratory Bird Conservation Fund. The 2023 addition brings the total protected area at the Neches River National Wildlife Refuge to over ten thousand acres.

An avid outdoorswoman, Julie helped improve access to the Neches River by establishing the first designated paddling trail along the river as part of the Texas Parks and Wildlife Department's nature tourism program. Julie is married to Cliff Shackelford, who served for thirty years at the Texas Parks and Wildlife Department as the state ornithologist; they have two children.

Janice's work has not been confined to the waterways of East Texas. She worked closely with parks advocate and director of the Texas Coalition for Conservation, George Bristol, to build the momentum to achieve a 2019 constitutional amendment to reinstate permanent funding for Texas State Parks and Texas Historic Sites through the Sporting Goods Sales Tax. In 2023 she served on an advisory team of the Texas Coalition for State Parks to secure $1 billion for the Centennial Parks Conservation Fund. The fund will be used exclusively to add new state parks.

Janice's enduring legacy can be found in the tens of thousands of acres of wildlife refuges, wildlife management areas, wilderness areas, and private preserves that exist because of her relentless efforts and ability to build coalitions. She also noted what is not there: reservoirs. "There are creeks, rivers, and lands that are not unnecessarily inundated."[30]

Janice's legacy also lives on in the work of the Texas Conservation Alliance. In 2021 the organization celebrated its fiftieth anniversary, hosting fifty events that included a thousand volunteers planting more than sixteen thousand trees and thirty-nine hundred native prairie plants. They also removed more than eleven thousand pounds of litter. In addition, volunteers constructed and installed forty-five wood duck boxes along the Neches River.

Under Janice's leadership, the Texas Conservation Alliance is continuing its twenty-year battle to oppose the creation of the Marvin Nichols Reservoir and save the Sulphur River in Northeast Texas. The proposed dam would permanently flood sixty-six thousand acres of rich, forested bottomlands, wetlands, and agricultural lands. The reservoir would displace thousands of residents from their homes and businesses.[31] With a shift in climate and weather patterns, Texas is predicted to experience intense droughts punctuated by extreme weather events that cause flooding. Those types of events will give water districts an impetus to call for more reservoirs. Evaporation during prolonged droughts continues to be an issue for all surface water reservoirs.

Perhaps most of all, Janice's legacy lives on in her son, David Bezanson. Following in his mother's footsteps, David has enjoyed a long career in conservation.[32] David currently serves as land protection strategy director with The Nature Conservancy's Texas Chapter. During his career he has focused on protecting preserves and private lands, with special focus on forests, native prairies, and the Texas Hill Country. He has worked to conserve additions to the Caddo Lake Wildlife Management Area and more than ten thousand acres in the Longleaf Ridge area, part of the one million acres divested by Temple-Inland. These accomplishments reflect the family's roots in East and Central Texas; some are places David visited as a child with his parents and grandparents.

David shared, "I was lucky to grow up with examples like Ned Fritz, and my mother, from a young age and it has been more inspiring than I can begin to convey. Because of my parents and Ned, I've never had any doubt about what is important to accomplish."[33]

In 2012 Janice and fifteen other women were chosen by the Bob Bullock Museum to be part of its Women Shaping Texas in the 20th Century exhibition. Janice has won a long list of awards, including being honored with the Audubon Texas Women in Conservation Terry Hershey Award in 2016. Each award represents her tenacity to save some of Texas' most scenic and wild places for future generations.

HONORING FAMILY HERITAGE THROUGH A GIFT OF LAND

Dessor Ree Frazier (1918–2004)

Dessor Ree Frazier embodies the idea that we all can do something for conservation on a scale that is meaningful to each person. Dessor honored her family and their connection to the land by donating twenty acres to the Texas Land Conservancy as the Frazier-Runnels Wildlife Preserve, which is part of the larger Angelina National Forest and a short distance from the Upland Island Wilderness.

The Angelina National Forest was established in 1938 in an effort to conserve some of the last remnants of the mixed-pine forests of Texas. (The Angelina National Forest was named for a Native American woman who helped guide Spanish explorers through the East Texas wilderness.) The woods contained in the Upland Island Wilderness are said to be some of the most ecologically diverse in the state, with more than 450 plant species and eleven unique habitats that support 170 species of birds, including the endangered Red-cockaded Woodpecker.

Dessor, her parents, and many of their extended family members are buried at the Rockwall Union Cemetery, which is surrounded by the wilderness area. Next to the cemetery is a marker designating the edge of the Catahoula Preserve, which was purchased by the Texas Land Conservancy in 1984 under the direction of its founder, intrepid Dallas attorney Edward C. (Ned) Fritz (1916–2008). The Catahoula Preserve, an inholding in the national forest, was the first property that the conservancy purchased.

The Texas Land Conservancy was founded in 1982 by Ned, known by many as the Father of Texas Conservation. In response to the increasing demand on land and water for development, Ned realized there was a need for a statewide land trust that would work to preserve lands both large and small, public and private. Ned gathered a group of grassroots volunteers to establish the Natural Area Preservation Association, now

Dessor Ree Frazier signing the documents to create the Frazier-Runnels Wildlife Preserve, 2001. Courtesy The History Center, Diboll, Texas

Members of the Frazier-Runnels family at the dedication of the plaque marking the Frazier-Runnels Wildlife Preserve, 2002. Courtesy David Bezanson

the Texas Land Conservancy, which currently conserves over 150 properties totaling more than 120,000 acres in Texas.[34]

Dessor gave the land in memory of her family, several of whom were born into slavery. Her great-grandfather James A. Frazier was born enslaved, later serving in the "colored" artillery battalion of the Confederate army during the Civil War. After emancipation, he moved his family to Texas to settle in one of the freedmen's settlements near Jasper and, later, Boykin.[35] Dessor's grandfather William T. Frazier was also born enslaved; later he bought the farm where Dessor grew up. The freedmen's settlements (sometimes called colonies) allowed African American families to form communities and "seek security in a climate of racial terror."[36] Laws in Texas and across the South made it difficult for people of color to purchase and retain land. Places such as the Boykin Settlement created safe communities where African American families could share resources to farm, accumulate land, start businesses, and educate their children outside of hostile, segregated white communities.[37]

Dessor's grandparents and parents were able to expand their land holdings to around three hundred acres. Even though farming was the primary livelihood for the family, Dessor's father, Harrison Frazier Sr., held other jobs to supplement the family's income. As a *Beaumont*

Enterprise article about Dessor's gift noted, "Harrison worked full time elsewhere. Some years he rode logs downriver for lumber companies. Others, he extracted stone from inside the Blue Hole near their home that was destined for rebuilding the Galveston seawall following the Great Storm of 1900."[38]

The same article went on to include a statement from Dessor about growing up on the farm: "'My grandfather was a Runnels and my father was a Frazier, and, yes, sir, I plowed. I plowed all day, with two mules,' acknowledged the petite 83-year-old to the small crowd before her. Dessor Ree Frazier stood next to the plaque [commemorating the gift of land], alongside the dirt road she used to walk each day to school two miles away and looked pleased at all the attention being lavished on her."

Dessor was born on the farm along with her three brothers and nine sisters. From an early age, she had a special connection with the land. Even though farming was not her first career choice, she later returned to it in retirement. Dessor saved enough money working six days a week as

Dessor Ree Frazier with Richard Donovan, surrounded by family and the conservation community, at the dedication of the plaque and land of the Frazier-Runnels Wildlife Preserve, 2002. Courtesy David Bezanson

a maid to attend Prairie View A&M University's nursing school. However, she found that segregation applied to even the highly educated, as she was denied employment at the Angelina County Hospital in Lufkin simply based on the color of her skin. She finally gained acceptance at the University of Texas Medical Branch in Galveston, where she worked her way up the ladder to become associate director of nursing. She worked there for thirty-eight years. She never married and never had children. After retirement, Dessor returned to the farm and her sister, the youngest of thirteen children, reported that Dessor raised cattle that she treated more as pets than as livestock.[39]

Her giving a gift of twenty acres of land from a farm that was difficult to acquire and maintain over multiple generations makes it even more meaningful. When she entrusted it to the Texas Land Conservancy, she stipulated that it be named for both sides of her family—Frazier and Runnels—and that it be managed "so future generations of Americans might come here and enjoy this wild place and wildlife that lives here."[40] It was also her desire to inspire others in her community to donate what they could from their farms to conserve the forest they explored and grew up with.

Black-capped Vireo, May 28, 2023. Courtesy Gil Eckrich

individual issues. These women exhibited awe-inspiring collective organizational and problem-solving abilities as they faced down opposition from other elected officials, developers, media, and sometimes their own neighbors.

THE SPRITS OF THE SPRINGS AND THE FIGHT TO PROTECT THE EDWARDS AQUIFER

The journey to establish one of the largest state parks in an urban setting began in earnest in the 1960s, culminating in 2005 with the opening of Government Canyon State Natural Area on the outskirts of San Antonio. However, the causes that led to the creation of the natural area go back over 120 years. While this might seem like a story about just setting aside land and habitat, the establishment of Government Canyon State Natural Area was part of a community-led, politically charged effort to conserve San Antonio's sole source of drinking water: the Edwards Aquifer.

Humans arrived in Central Texas around twelve thousand years ago. The early people followed the migrating herds and changing weather. One of the seminomadic societies was the Coahuiltecan. The creation story of the Coahuiltecan included the life-giving springs that flowed from the Balconies Fault, with the San Antonio Spring, otherwise known as the Blue Hole, at the center of their universe.

The spring is also the headwater of the San Antonio River and bubbles forth from an artesian well. The well and the springs that flow from the Balconies Uplift are believed to be included in the White Shaman Mural, painted between two thousand and four thousand years ago on the cliffs above the Lower Pecos River near Seminole Canyon State Park and Historic Site. The mural depicts what is considered to be a map of the life-giving artesian springs, which includes San Antonio / San Pedro, Comal, San Marcos, and Barton springs.[1]

Fast forward a few thousand years to the Mission San Antonio de Valero, later known as the Alamo, which was established in 1718 near the San Antonio River and approximately three miles south of the springs of the Blue Hole.

For decades, all the drinking water for the citizens of San Antonio came from the river, surrounding creeks, or shallow wells. The river and creeks also served as the dumping ground for all manner of waste.

U. S. GEOLOGICAL SURVEY EIGHTEENTH ANNUAL REPORT PART II PL. XXXIX

ARTESIAN WELL, CITY WATERWORKS COMPANY, SAN ANTONIO, TEXAS.

Artesian well, City Waterworks of San Antonio, circa 1896. US Geological Survey, eighteenth report, public domain

Mary Eleanor Brackenridge, date unknown. Courtesy University of Texas–San Antonio Special Collections

In 1869 banker, businessman, and philanthropist George W. Brackenridge purchased the land that included the San Antonio Springs and the Blue Hole. George lived with his mother and sister, Eleanor, in a large Victorian-era home that he constructed on the bluff above the headwaters. Mary Eleanor Brackenridge (1837–1924) was also a banker, philanthropist, and active member of the Federation of Women's Clubs. She and her brother both supported the suffrage movement and the concept that women could—and should—work outside the home. Both brother and sister valued education for everyone, including women and people of color, and funded scholarship programs and schools to improve the lives of the underserved.

Concerned about sanitary drinking water, the Brackenridges decided it was time for San Antonio to have a more efficient water system. They

Sister Mary Gabriel, Mother Mary John A. Flasseur, and unknown woman enjoying the headwaters of the San Antonio River near the Blue Hole, date unknown. Courtesy Sisters of Charity of the Incarnate Word Archives, San Antonio, Texas

drilled several wells into the artesian layer that rims the edge of the deeper aquifer. From that point on, water was big business in San Antonio. The population in San Antonio in 1890 was around eighteen thousand people, and it swelled to nearly fifty-three thousand by 1900. Taking advantage of the clean water, several breweries opened, including what would later become the Pearl Brewery.[2] Even with the addition of clean drinking water, sanitation remained an issue, and waves of diseases such as yellow fever and cholera swept the city.

The Sisters of Charity of the Incarnate Word traveled from France to San Antonio in the 1890s to assist during one of the more intense cholera outbreaks. George Brackenridge was ready to sell his land that contained the Blue Hole and chose to sell it to the Sisters of Charity. In 1897 the sisters purchased the land, eventually established the University of the Incarnate Word, and carefully watched over the sacred spring for more than 120 years.[3]

Between World War I and World War II, the population surged again. Even with the addition of wastewater treatment plants located on the south side of the county, the City of San Antonio wrestled with how to adequately supply clean drinking water and wastewater services to all citizens. The wastewater system relied on gravity flow, which took waste and everything else south of the city via creeks and rivers. During flooding events, the wastewater treatment plants at Mitchell Lake often leaked out of the containment ponds and into the Medina River. However, since the south side was primarily home to nonwhite, socioeconomically disadvantaged communities, not much was done.

Fay Bloom (1918–2009) was born and raised in New York City, where her father worked for William Randolph Hearst in the newspaper business. It was not until she attended Randolph-Macon Women's College in Lynchburg, Virginia, that she first witnessed "the racial tensions between whites and African Americans and became conscious of her own identity as a Jewish woman." It was in college during the 1930s that she observed a truck filled with African American children flip over, and no white person would help them. She felt powerless and angry as the white men stood and watched while the children suffered.[4]

In 1942 she agreed to go on a blind date with a friend of the family, Bill Sinkin, who was visiting from Texas. Bill didn't even make it through the date before proposing to Fay, to which she responded, "You guys in Texas must be hard up."[5]

Bill made a good impression, and they married in May 1942. The newlyweds returned to Bill's home in San Antonio, where he owned and operated a dry goods store. After arriving in San Antonio, Fay became aware of—and was appalled by—the sanitation conditions of the city, especially in the neighborhoods that were majority African American or Mexican American. The residents in those neighborhoods primarily had

Privy toilet with child at the common-use freshwater pump, Mexican American rowhouse, 1939. Photo by Lee Russell. Farm Security Administration—Office of War Information Photograph Collection, Library of Congress Prints and Photographs Division, Washington, DC

privy toilets or outhouses that were dug directly into the earth or employed buckets that were dumped into the rivers and creeks. During heavy rain events, the raw sewage turned into a toxic sludge that covered people's yards, gardens, and streets.

In 1947, just a year after the birth of her son, Fay became president of the League of Women Voters. Sanitation and access to clean water in underserved neighborhoods became her primary focus. She and the League of Women Voters advocated and pushed the city council to extend proper infrastructure to all citizens, not just the ones in middle-class or wealthy white areas. An article in the *San Antonio Express-News* states, "For women like Sinkin, the league was an important outlet because there was little opportunity for them to participate on boards or in institutions that made policy. In those years, for example, women couldn't serve on juries or grand juries. That didn't happen in Texas until 1957, and Sinkin and the late Kathleen Voigt were the first women in San Antonio appointed to a grand jury. 'Locally, women played a very insignificant role because men were so macho,' Sinkin recalled in a 1990 interview with the *San Antonio Light*. 'I remember going to the City

Woman and children at Mexican American rowhouses, 1939. Photo by Lee Russell. Farm Security Administration—Office of War Information Photograph Collection, Library of Congress Prints and Photographs Division, Washington, DC

Council to plead with them for a sanitation engineer because our public health standards were so bad. The mayor said I should go home and cook my husband's dinner.' "[6]

Fay's efforts to improve sanitation and her service as the first president of the Visiting Nurse Association won her an appointment as the first woman to serve on the San Antonio Health Board. Shortly after the Supreme Court ruled segregation of schools was unconstitutional, Fay ran for the school board in San Antonio. She lost but went on to advocate for changes such as having a dietitian for the school district and advanced classes for students; for that, she was deemed a Communist.[7]

The same year Fay ran for the school board, the state was gripped by one of the worst droughts in history, and the wells in San Antonio were drying up, as were the rivers and creeks. In 1956 Comal Springs in New Braunfels stopped flowing for more than one hundred days. The lack of flow decimated one of the small fish that later would become protected under the Endangered Species Act.

In 1959, as a response to the drought and lack of flow, the Texas legislature created the Edwards Underground Water District. The purpose of the district was to "conserve, preserve, and protect" the Edwards Aquifer. The Edwards Aquifer was defined as covering about thirty-six hundred square miles in Kinney, Uvalde, Zavala, Media, Frio, Atascosa, Bexar, Comal, Guadalupe, and Hays Counties, with an even more expansive recharge zone. In typical fashion, the legislature failed to grant the water district any type of regulatory power.[8]

Compounding the drought was the post–World War II explosive growth across Central Texas. In the late 1950s and 1960s, an inner highway loop known as Interstate 410 was constructed to encircle the city of San Antonio. By the late 1960s and early 1970s, another ninety-five-mile highway called Loop 1604 was being constructed north of Interstate 410. The new highway extended beyond the city limits and, in many places, pushed through unincorporated country. Both new highways cut through the recharge zone of the Edwards Aquifer.

The roads opened up development opportunities. One of those projects was the University of Texas–San Antonio campus, which was built near the intersection of Interstate 10 and Loop 1604 just north of Leon Creek. The campus sparked controversy as some opposed its construction for being too removed from the city center, while others argued that a development of that size should not be built over the recharge zone.

By 1970 the population of San Antonio had grown to more than 654,000 people, not including the surrounding communities, and was set to grow even more as military personnel returned from the war in Vietnam.

In 1973 the Endangered Species Act became law after scientists found that multiple species of "fish, wildlife, and plants in the United States have been rendered extinct as a consequence of economic growth and development unhampered by adequate concern and conservation." The act set in motion a means to protect those species in peril.[9] In Texas and around the country, biologists were already making the case for which species and their ecosystems were in need of protection. At least one salamander and a fish species that exists only in the Comal and San Marcos Springs were already considered endangered, with several more being monitored.

In 1975, under the newly passed Safe Drinking Water Act, the US Environmental Protection Agency designated the Edwards Aquifer the sole source of drinking water for San Antonio. At that point, it was becoming apparent that something had to be done to save the aquifer.

Fay Sinkin and the League of Women Voters stepped up to the plate to form the Aquifer Protection Association. The association's goals included raising awareness about water conservation and purchasing nearly twenty thousand acres of land over the recharge zone as a Bicentennial project. The group circulated a petition to acquire more than one hundred thousand signatures to pressure local, state, and federal governments to fund the purchases.[10] Part of the issue with that early plan was that a Texas law prohibited cities from using tax dollars to purchase lands outside the city limits. Unfortunately, much of the recharge zone existed just outside the city limits in Bexar and Comal Counties.

With the opening of the highways, developers began buying land in the northern part of Bexar County at an alarming rate. Hundreds of applications for zoning changes and variances for retail stores, laundromats, automotive centers, restaurants, bars, apartments, office buildings, light industry, and shopping malls were submitted each week. One shopping mall that was planned to be built directly over the recharge zone at the corner of Loop 1604 and Highway 281 sought to cover 129 acres with buildings and parking.[11]

North Star Mall being built near Loop 410 and San Pedro Avenue over the Edwards Aquifer, circa 1968. Courtesy University of Texas–San Antonio, Libraries Special Collections

The shopping mall again sparked controversy, and Fay and the Aquifer Protection Association set out to obtain the signatures they needed to place a referendum on the ballot to overturn the zoning privilege the mall was given. (They needed twenty-seven thousand signatures—they collected forty-nine thousand.) The measure made it onto the ballot, and citizens voted for the referendum, stopping the mall and development over the recharge zone that was within city limits.

The bulldozers did not sit idle for long, as the courts later proved that the Texas Constitution did not allow cities to utilize referendums at that time. However, the city council voted to honor the referendum and reversed their votes on allowing the mall.[12] The referendum opened the eyes of the citizens of San Antonio and empowered them. It also opened the public's eyes to how "business as usual" at city hall was stacked in favor of developers, and not always in the best interests of the health and safety of the people. For example, at the time, the president of the Greater

Fay Sinkin delivering petition signatures to the San Antonio City Council, November 1975. Courtesy University of Texas–San Antonio, Libraries Special Collections

San Antonio Builders Association was also the chair of the City Zoning and Planning Commission.[13]

While Fay and the association were fighting to protect the inground—and rechargeable—aquifer on the north side of San Antonio and Bexar County, the city was considering adding a surface water reservoir on the south side. The proposed Applewhite Reservoir would flood thousands of acres of land and displace families in one of the most disadvantaged communities. It would cost taxpayers more than $100 million, not including the cost of new pipelines and pumping stations.

After approving the idea, the city council used eminent domain to secure the thousands of acres planned for impoundment. However, questions still remained. How did San Antonio have the right to the water? How would the city ensure that farmers and landowners downstream who already claimed rights would receive their allotted share? In the 1971 *Report on Reclamation and Re-Use of Municipal Wastewater*, the idea of refilling downstream water supply with treated wastewater was reviewed in depth.[14] The treated water would flow from the three wastewater treatment plants into the cooling lakes of two coal-powered electrical plants and eventually be released into the southern reaches of the Medina River to be resold. That formula would allow the city to capture fresh water upstream in the spring-fed Medina River for drinking water. Not everyone thought the formula was wonderful.

Maria Antonietta Berriozabal did not think water would become the centerpiece of her public service when she decided to run for City Council District 1 in 1980. She won the election and became the first Mexican American woman to serve on the council, and one of the first Mexican American women in the state to be elected to any office. She was the elected ally the water conservation movement needed.[15]

Maria Antonietta Berriozabal campaign button, date unknown. Courtesy Our Lady of the Lake University Center for Mexican American Studies and Research Special Collections

Maria's grandparents moved to Laredo in 1910 from Mexico during the Mexican Revolution, and her parents moved to San Antonio in 1942 shortly after Maria was born. After graduating from high school, Maria worked to supplement her parents' income and support her siblings. She was in her thirties when she began night classes to earn a political science degree from the University of Texas–San Antonio. She worked several jobs with the Democratic Party, until she eventually interned for Henry Cisneros, who was, at the time, a council member and eventual mayor, and later became the secretary of housing and urban development under President Bill Clinton. While working for the Democratic Party and advocating for public health issues, Maria met Fay and Bill Sinkin.

In an interview with David Todd in 2006, Maria shared,

> When I got elected to the City Council, I didn't say to myself, "Water is going to be the number-one issue that I'm going to deal with." I was dealing with housing and jobs and transportation and garbage and potholes and the neighbor next door and the dog barking. . . . And then, water came up, and what I remember was that when I was little, we used to go to my grandfather's farm. They had a well, and my uncle would take us to the well. He would take a pail that was made out of wood, drop it down the well, and you could hear when it hit the water. I remember the splash, and then he would haul it up. . . . I remember that water, how it tasted. . . . I have a picture of and a taste for water. So, when I hear that there's benzene in the aquifer, when I hear that they have to close some wells because the children are getting sick, it's so immediate. It's an emotional response and even a physical one.[16]

In the same interview, Maria reflected that a study revealed that the soils in and along Leon Creek and the Medina River that were to be flooded for Applewhite Reservoir were contaminated with toxins. Despite that information, the city still voted to build it. Maria and councilwoman Helen K. Dutmer (1920–2020) were often the only council members to stand against the project.

Helen was elected to the city council in 1977 from the southside district. She served fourteen years on the council and went on to become the

City of San Antonio Councilwoman Helen Dutmer, 1979. Courtesy University of Texas–San Antonio, Libraries Special Collections

first woman elected as a Bexar County commissioner in 1991. In addition to raising her daughter and working at the family business, Helen also served on the San Antonio Planning Commission, Mission Neighborhood Council, Mission Task Force, San Antonio Conservation Society, and board of the San Antonio Mission National Park.[17]

Fay, Maria, Helen, and many others feared the reservoir would allow for unfettered building over the aquifer. Heeding the call from Fay and other emerging environmental groups, Maria also began to see the correlation and the value in protecting the aquifer from development. Each new development over the environmentally sensitive recharge zone had the potential of contaminating the water from leaking sewage pipes (something San Antonio had plenty of) or from oil, gasoline, and antifreeze washing off parking lots and driveways during rainstorms into the soil, creeks, and deep underground fissures and caverns that retained the water. Maria stated, "For the ten years that I was on the City Council, I never once voted for a zoning case to permit building over the aquifer because, at the time, developers were coming almost every week with zoning cases that would build over the water."[18]

City of San Antonio Councilwoman Maria Antonietta Berriozabal, date unknown. Courtesy Our Lady of the Lake University Center for Mexican American Studies and Research Special Collections

Fay ran for a position on the Edwards Underground Water District board in 1983. She beat nine other candidates for the coveted position and became the first woman to serve on the board. Fay advocated for a management plan to curtail building over the recharge zone and put limits on pumping water and flows so everyone could have their "fair share." In a meeting of the board, the representative from Uvalde, Rodney E. Reagan, retorted, "My fair share is to get all I can capture."[19]

By 1987 Fay, Maria, Helen, and other concerned citizens were advocating to have the city identify the sensitive areas of the recharge zone to prohibit development in those areas and encourage development in less sensitive ones. The idea was to treat the recharge area as a reservoir that happened to be underground instead of aboveground. Just as development could not happen on a surface reservoir, why not limit development on top of the underground lake? The women hosted community hearings and events to inform the public and listen to their concerns.[20] Each believed strongly in communicating regularly with the citizens in their

Fay Sinkin giving a presentation about the Edwards Aquifer Recharge Zone, 1974. Courtesy University of Texas–San Antonio, Libraries Special Collections

neighborhoods to keep them informed and engaged. It was a strategy that ultimately built trust in them and for the cause.

In 1988 Fay invited eight influential conservation leaders to sit around her kitchen table and brainstorm how they could buy and conserve lands over the recharge zone. Fay and Maria both called for studies to be conducted. That same year, one of the groundwater studies from the US Geological Survey revealed that several of the city's wells contained fecal matter, fertilizer, and petroleum-based toxins. The findings pushed the council to begin considering stricter zoning and enforcement over the recharge and surrounding buffer zones.

Meanwhile, the fight for and against the Applewhite Reservoir continued. The discovery of archaeological sites had slowed momentum considerably, but the plan was still moving forward at the expense of taxpayers. In 1990 the San Antonio Conservation Society and Southern Texas Archaeological Association sponsored a public meeting with the US Army Corps of Engineers and Councilwoman Maria Berriozabal. The public hearing addressed some of the issues, including the multiple lawsuits filed by landowners adjacent to the reservoir. Maria made the case for scrapping the project altogether.[21]

By 1991 the taxpayers and most of the city council had had it with Applewhite. Mayor Lila May Banks Cockrell (1922–2019) had at first supported the project, but even she was beginning to turn toward the idea of halting it.[22] When Lila first became mayor in 1977, she and the council had approved funding to move forward with the venture, which was originally approved in the 1950s. Lila served as president of the League of Women Voters, became the first woman mayor pro tem, and first woman mayor of San Antonio.

With the help of Kay Turner of the Homeowners-Taxpayers Association, a petition was circulated to gather signatures to place a referendum

Mayor Lila Cockrell (*front*) and unidentified woman at the dedication to make Mission Parkway a national trail, April 1977. Courtesy University of Texas–San Antonio, Libraries Special Collections

Lila Cockrell being sworn in to the San Antonio City Council, 1963. Courtesy University of Texas–San Antonio, Libraries Special Collections

on the ballot to put an end to the Applewhite project once and for all. In opposition, the city water board, under the leadership of Chair Nelda Weatherly, spent hundreds of thousands of dollars in advertising to try to convince voters to support the reservoir. The issue became a "hot potato" that was being watched across the state as other cities and counties were considering how to move forward with their own water projects. The referendum made it onto the ballot, and voters clearly wanted out from under the expensive development. A few days following the public vote, the city council also voted three to six to kill the reservoir. However, the city water board hotly contended that the referendum was illegal.[23]

San Antonio was trying to rebrand itself as a vacation destination in the late 1980s and early 1990s. In addition to making improvements to the River Walk along the San Antonio River, the city approved the development of Sea World on the northwest side of town just inside Loop 1604. The site sprawls across 250 acres and cost around $170 million to build.

The main attraction, the killer whales, reside in a seven-million-gallon pool, with smaller pools for other species. The water-heavy project enraged environmentalists and citizens fighting for their right to clean, abundant water. When it opened in 1988, Sea World was said to be a huge success because it brought vast numbers of tourists to town. With each visitor came the need for more hotels. More hotels meant more toilets and taps. More flights to the airport built on the edge of the aquifer. More convenience and retail stores. More water use across the board.

Sea World's construction was followed quickly by the approval to build Six Flags Fiesta Texas in an old limestone quarry just a few miles away. But it was the PGA golf course and resort hotel planned to be built directly over the recharge zone that garnered the most attention. In 1984 Lumbermen's Investment Corporation submitted plans to build a thirty-six-hole PGA golf course, a five-hundred-room hotel, and luxury homes on a twenty-eight-hundred-acre piece of land just north of Loop 1604 and Highway 281. And then, the Texas boom went bust in the mid-1980s. The project was paused until 1994, when Lumbermen's and twenty-six other developers banded together to exploit a loophole in the water quality ordinance that the city had passed during the years the project was dormant.

In 1991 the Sierra Club filed a lawsuit against the US Fish and Wildlife Service, citing negligence to provide necessary protection and adequate spring flow required by the Endangered Species Act. US District Judge Lucius Bunton ruled in favor of the Sierra Club. The lawsuit triggered a series of regulations to ensure spring flow and required the Texas legislature to control withdrawals from the Edwards Aquifer. Senate Bill 1477 established the Edwards Aquifer Authority and dissolved the Edwards Underground Water District. However, in the establishment, the legislature violated the Voting Rights Act because the board was appointed rather than elected, so its ability to make progress was delayed until 1995.[24]

Meanwhile, thousands of acres of land and property deeds were piling up in the Resolution Trust Corporation as they were lost in bankruptcies. The Resolution Trust Corporation was established in 1989 by the Financial Institutions Reform Recovery and Enforcement Act to absorb the defaulted loans and assets of the savings and loans and other banks that had lent to "anyone with a pulse." Between 1989 and 1995, the Resolution

Karen and Tim Hixon at the grand opening of Government Canyon State Natural Area, 2005. Courtesy Texas Parks and Wildlife Department

Sinkin passed away, but not before seeing the city transform from a place where underserved populations lacked access to clean water, parks, or adequate health care to a community that fiercely protected its sole source of water and embraced parks as part of its healthy-living strategy. Fay's legacy lives on in her sons, but it can also be found in every clean drop of water a citizen or visitor of Bexar County drinks and in the thousands of acres of lands preserved. Just two months after Fay's passing, the City of San Antonio transferred three thousand acres to Government Canyon State Natural Area.[37]

Today, the twelve-thousand-acre day-use-only park offers hiking and biking trails and a natural play area mostly constructed through funds raised by the Friends of Government Canyon State Natural Area. When the endangered Golden-cheeked Warblers are in residence for the nesting season, a portion of the park is closed to limit interface between the birds and humans. The visitor center is dedicated to Tim and Karen Hixon, and a plaque nearby recognizes the value of Fay Sinkin's efforts. When Tim completed his six-year term on the Texas Parks and Wildlife Commission, Karen was appointed to it in 2007 by Governor Rick Perry.

From left: Councilwoman Delicia Herrea, Bill Sinkin, and Superintendent Dierdre Hisler at the grand opening of Government Canyon State Natural Area, 2005. Courtesy Texas Parks and Wildlife Department

She continued her conservation journey through serving on the boards of the Texas Parks and Wildlife Foundation, Texas Nature Conservancy, and San Antonio Zoological Society to name a few.

Construction to widen Highway 281, which cuts north from Bexar County to Comal County and beyond, started in 2020. The increased carrying capacity has spurred a new surge of development and speculation across the recharge zone, yet again increasing the price per acre and making it more challenging than ever to conserve lands and create parks. According to the 2020 census, the population in the ten counties that contribute to the aquifer now exceeds 2.6 million people, with 1.9 million living in Bexar County alone. This increase in population strains the municipalities' plans to provide parks for people, enforce water conservation, and meet the desire to grow their respective economies. One answer lies in the public-private partnership effort demonstrated in the successful establishment of Government Canyon State Natural Area.

Coming full circle, in 2020 the Sisters of Charity of the Incarnate Word signed a conservation easement with Green Space Alliance of South Texas to preserve the fifty acres around the blue hole and the headwaters of the San Antonio River.

Thanks to the vision of Deborah Morin and longtime water advocate Dianna Wassenich, the life-giving springs that were the heart of the Coahuiltecan creation story will continue to be honored and connected via the Great Springs Project. In 2019, under the direction of Garry Merritt, the Great Spring Project seeks to again link the springs via a trail system that begins its journey at the Blue Hole in San Antonio and terminates at Barton Springs in Austin. The public-private partnership endeavor also secures conservation lands, parks, and culturally significant spaces as it links the communities along the network of springs. With construction already well underway, the entire one-hundred-mile system is expected to be completed by the Texas Bicentennial in 2036.[38]

Persistent drought from 2023 to 2025 resulted in Median Lake, just ten miles from Government Canyon State Natural Area as the crow flies, to fall below two percent full—otherwise known as "dead pool" level. At the same time, the Edwards Aquifer reached historic low levels prompting stage 5 water restrictions. The drought has yet again engaged the communities along the Edward's Aquifer in rethinking water use and conservation.

PROTECTING THE ENVIRONMENT THROUGH POLICY AND PLANTS

Susan Rieff (1955–2023)

Born and raised in Fayetteville, Arkansas, Susan Rieff participated in student government, orchestra, cheerleading, and the honor society in high school. Like many young people her age, her mind was awakened to environmental issues that plagued the country: "On the first Earth Day in 1970, we learned that 40 percent of the country's water was contaminated. We were becoming aware that nature was not perfect, that overpopulation was a problem. I first heard the concept of limited growth around that time. The country was consuming more than we were producing. That can't be good."[39]

After being awarded a scholarship, Susan came to Fort Worth, Texas, in 1973 to attend the newly opened Department of Environmental Science at Texas Christian University. She went on to earn a master's degree in public policy from the University of Texas LBJ School of Public Affairs and later earned her doctorate from Duke University in North Carolina. While at the LBJ School, she was selected for a Presidential Management Fellows Program internship. The internship allowed her to test various positions and issues within the Department of the Interior and other agencies, including the office of Senator Dale Bumpers of Arkansas. The senator served on the Public Lands Subcommittee. At the end of the four-month internship, Senator Bumpers offered her a position as a legislative aide, with a focus on the environment and natural resources.[40]

"I worked for Senator Bumpers when Jim Watt was secretary of the Interior (he was such a terror about protecting the resources of the department), and Senator Bumpers was his chief antagonist," said Susan. "We worked to hold off some of the initiatives of the Reagan administration, such as stopping Secretary Watt from giving away federal coal resources at very low prices."[41] James Watt, as secretary of the interior under President Ronald Reagan, had a personal mission to sell national parks and public lands and open them to mining and oil drilling. Thanks to Senator

Susan Rieff and Senator Dale Bumpers, 1985. Susan Rieff personal collection

Bumpers and his team, a moratorium was placed on coal leasing for a time.

Susan returned to Texas in 1985 to take a newly created position at the Texas Parks and Wildlife Department. Texas Parks and Wildlife Commission Chair Ed Cox Jr., a Dallas businessman in the oil and gas industry, wanted the agency to take a greater role in protecting the environment. He valued healthy habitats to strengthen the agency's hunting, fishing, and recreation missions. Along with fellow commissioner Bob Armstrong, Ed Cox Jr. called for the creation of the Resource Protection Division. When Susan took the position as its first director, she also became the first female division director at the agency.

Susan was twenty-nine years old when she walked through the doors of the Texas Parks and Wildlife Department in Austin. After a quick tour of the agency's new facility, she was shown to her desk—not an office, just a desk sitting in the hallway. The commission saw the need for the new division; however, the agency's executive director, Dickie Travis, was less enthusiastic about the new focus on resource protection.[42]

The desk in the hallway could be viewed as a metaphor for the fact that certain people in the agency saw the division as temporary. Susan eventually received an office and was able to hire a staff of twenty-five to serve as the state's voice for the environment. She and her team worked on projects both small and large, including disseminating information about why deepening and expanding the Houston ship channel would not be wise. The topic was controversial and involved almost every environmentally focused organization in the state, as well as many national ones. Susan often stated that many of the projects she worked on were not popular with the legislature or certain industries.

After three years creating and staffing the new division, Susan changed positions to become the assistant commissioner under Jim Hightower at the Texas Department of Agriculture. Jim Hightower had progressive ideas that often clashed with those of the Texas Farm Bureau and less progressive farmers. While at the Department of Agriculture, Susan focused primarily on regulating pesticides and incentivizing organic and more holistic food production.

The pesticide issue was pervasive, and in 1989 a report from West Texas showed that of the 116 wells tested, forty-four were contaminated by agricultural pesticides. According to the report, "This round of tests appears to confirm the pattern shown in our earlier testing that pesticides can—and are—contaminating our groundwater resources in Texas." In those tests, concerning levels of arsenic, nitrates, bromacil, and atrazine were discovered in the wells in the nine-county sampling area.[43]

The intent of the testing was to collect data to show how these chemicals remain in the soil and water long after application and what long-term effects on humans, plants, wildlife, and aquatic species they have. Transitioning to organic methods and rethinking the use and application of pesticides were hard sells since the agency and agriculture industry had championed better living through chemicals for decades. Susan and Commissioner Hightower's team started the conversations and promoted the more environmentally and human-friendly alternatives.

In 1990 State Treasurer Ann Richards was running for governor against businessman Clayton Williams of Dallas. It was a tight race until March, when Clayton made a comment that became one of many factors that changed the course of the election. Clayton was holding a press event at his ranch when it started to rain; he then commented, "Rape is kinda like the weather. If it's inevitable, relax and enjoy it." He later apologized for the remark, but by then it had become a hot topic—even at the national level.[44] Ann won the election to become the second woman governor in Texas and the last Democratic governor for more than thirty-five years.

Governor Ann Richards recruited a large number of talented women to serve in her administration, including Susan Rieff, whom she hired as director of environmental policy and legislative liaison. When Ann was a Travis County commissioner, she had supported several environmental issues that were surfacing, especially finding solutions to protect the eight endangered species that resided in her county.

While Ann was governor, Texas was at the forefront of several major environmental issues and policy shifts, including better enforcement of the Endangered Species Act. The US Fish and Wildlife Service had published an assessment of the endangered species found across the state and, following procedure, opened a public comment period. A newspaper in San Antonio "published a map of what critical habitat might look like in Central Texas for the Golden-cheeked Warblers," Susan remembered, "and all hell broke loose because thirty-three counties were listed on the map, and the spin was that it would shut down development in those thirty-three counties, and people's lands would be taken away. It launched the Take Back Texas private property wars. . . . Every politician in the state was embroiled in this because of the private property rights reaction to it. . . . It was brutal."[45]

In addition to the Endangered Species Act, Susan had to find solutions for the newly amended Clean Air Act as well as wrestle with policy and pollution issues. Those issues ran the gamut from major source-point polluters such as Formosa Plastics Corporation along Lavaca Bay, to new auto emissions standards, to encouragement of counties to pave roads to reduce dust. Clean air policies and actions were popular with citizens, especially those living downwind of some of the larger coal and plastic manufacturing plants, but they were not popular with industry. Governor Richards appointed Kirk Watson as the first chair of the Air Control Board, and Texas became the first state to pass new regulations to comply with the Clean Air Act. Her administration also passed legislation to clean up the bays and estuaries along the coast.[46]

Susan also faced an issue with companies using Texas as a dumping ground for hazardous waste, a controversy that came to a head when a New York company sent a trainload of toxic sludge to Texas for disposal. With Susan's assistance, Governor Richards introduced a bill to place a moratorium on locating new hazardous waste sites in Texas and reduce the formation of the types of hazardous waste being piled into those sites. Susan admitted the bill was greatly watered down by the time it passed the Texas legislature; however, it was a start.[47]

Part of the problem with tracking and regulating pollution and waste was that various state agencies oversaw different aspects of the environment. For example, before merging in 1977 to form the Texas Department of Water Resources, the Texas Water Rights Commission, Texas Water Development Board, and Texas Water Quality Board each administered a particular aspect of water use. Susan recalled her experience bringing together the various agencies to strategize how to create a single oversight commission: "We had this huge meeting with all the people from the different agencies, who weren't necessarily keen on the idea because, of course, they had their own structures and their own levels of authority and their own budgets. . . . And so, we really started with a proposal to just bring the Air Control Board, Water Commission, and solid waste functions from the health department together. That was finally achieved."[48]

Eventually, the many agencies were brought together under the Texas Natural Resource Conservation Commission, which formed in 1993. The conservation commission purposely did not include the word *environment* in its name, as certain demographics had demonized the word, and insisted on using *conservation* instead. It wasn't until 2002 that the commission's name changed to the Texas Commission on Environmental Quality.[49]

Governor Richards lost the election in 1994 to George W. Bush. With Ann's term complete, Susan began a position as the deputy chief of staff for Interior Secretary Bruce Babbitt in the Clinton administration. While that position took her all over the country, it often landed her back in Texas. In 1996 the Barton Springs salamander was being considered for listing as an endangered species after President Clinton ended a moratorium (that was initiated through a bill sponsored by US Senator Kay Bailey Hutchison of Texas) on all new listings. When the moratorium was lifted, more than two hundred species were eligible for consideration. The sheer number caused a delay in many of the biological surveys and data required to decide their status.[50] The salamander's listing offered further protections for the Barton Creek watershed in Austin, as the salamander cannot survive in polluted or contaminated water.

After her service with the Department of the Interior, Susan went to work as the National Wildlife Federation's regional vice president and, later, as director of its lands program. The position reconnected her with an Austin project through the creation of the Balcones Canyonlands National Wildlife Refuge. Just as the refuge was about to be fully funded, Congress reduced the amount that had been appropriated. Multiple willing landowners were ready to sell, but no money was available to finalize the transactions. "The vocal critics of the Endangered Species Act are always saying that if the federal government wants to protect endangered species, it ought to buy the land. Here you have a situation where a federal agency wants to buy it, and someone is intervening to say no," said Susan in a 1999 *Austin American-Statesman* article.[51]

The money eventually was approved, and the Balcones Canyonlands National Wildlife Refuge now spans twenty-seven thousand acres in Travis and Burnet Counties, protecting Golden-cheeked Warbler and Black-capped Vireo habitat as well as the sinkholes and caves that feed water into the aquifers and Colorado River.

In 2004 Susan became the director of the Lady Bird Johnson Wildflower Center. The decision to accept a job with less travel came after she was diagnosed with breast cancer in 2003. Treatment halted the cancer in her breast, but it later returned in her liver and bones—and, eventually, her brain. She fought like a prize fighter to eradicate the cancer and put it in remission. However, the treatment and disease greatly sapped her energy.

Despite the agony of her cancer and treatments, Susan oversaw the addition of several new buildings and gardens at the wildflower center and worked diligently with Board Chair Ellen Temple to transfer the wildflower center to the University of Texas system in 2006. The transfer had been under negotiation for many years but was given the final boost when Lady Bird Johnson herself gave her approval.

The merger allowed the wildflower center to have access to broader research resources, form new partnerships, and become a leader in the landscape architecture world.[52] Like others who have worked at the wildflower center, Susan saw herself as a steward of Lady Bird's vision. In a Texas Legacy Project interview, Susan stated, "I read a lot of what Lady

Former First Lady Laura Bush (*left*) and Susan Rieff at Lady Bird Johnson Gala, 2011. Photo by Ken Nordhauser. Courtesy Lady Bird Johnson Wildflower Center archives, University of Texas–Austin

Bird Johnson said in speeches and books. And she was really a farsighted, wise person. She was talking about sustainability twenty years ago. She was talking about global warming twenty, thirty years ago. And so what we've done with the Wildflower Center is try to sort of capture that larger vision."[53]

In 2014, after a multiyear capital campaign, the wildflower center opened the Luci and Ian Family Garden. The garden was the wildflower center's answer to a need to connect more children and families with nature and give them intentional spaces to do just that. The family garden features places to run under a waterfall, splash in a creek, pretend to be a bird in a giant nest, watch butterflies, construct a fort out of loose parts, or just sit quietly and enjoy the sounds of nature. All this was made possible by Lady Bird's daughter, Luci Baines Johnson, and her husband, Ian Turpin.

After ten years with the wildflower center, Susan accepted a position with the University of Texas Center for Sustainable Design before retiring in 2016. Her legacy is laced within the layers of policy she helped shape and foster at the national, state, and local levels. Texas is one of the few states with its own commission on environmental control. She also is to be thanked for what we do *not* have in Texas, including unchecked hazardous waste dumps. Susan was one of the women selected for the inaugural Audubon Texas Women in Conservation Terry Hershey Award in 2014. Sadly, she passed away just as this book was being completed in the summer of 2023.

ENDANGERED SPECIES, WATER, AND BONDS

Valarie Scott Bristol

Between 1987 and 1990 eight species were listed as endangered in Travis County and the city of Austin. All of the species existed in the rugged hills just west of the city's center. Habitat loss and pollution were cited as the major threats. The fight to protect them and the drinking water held in the aquifers and lakes began to heat up just as unprecedented growth hit the region.

Members of Travis Audubon had been raising the alarm for decades that the nesting habitat for the Golden-cheeked Warblers and Black-capped Vireos needed to be protected. Every Golden-cheeked Warbler is born in Central Texas and depends on mature Ashe junipers for its nesting materials and insect-rich live oaks as its food source.

As early as 1987, people like Barbara Dugelby with Earth First! were calling on city council to take action. "We cannot afford to play games with these species. We cannot afford to mitigate damages after habitats of threatened and endangered species have been destroyed," Barbara said in an *Austin American-Statesman* article in 1988.[54]

The issue that brought the most attention to the need to find community-based solutions quickly was a large-scale development near Lake Travis called Steiner Ranch. The project called for a widening of Ranch Road 620 and Comanche Trail. The development and road were in the heart of prime endangered species habitat. Earth First! and city council member Max Nofzinger called for a citizen advisory board to be formed to help generate solutions.[55] Meanwhile, Earth First! occupied several of the caves where the endangered cave species resided to protest development.

By the fall of 1988, Earth First! and the citizen advisory group called on the county to create a regional comprehensive plan. County commissioner of Precinct 2, Bruce Todd, and County Judge Bill Aleshire both saw the value in applying for a blanket 10-A permit from US Fish and Wildlife,

Barbara Dugelby with Earth First! making a speech at the Cactus Café, 1987. Courtesy Barbara Dugelby

which would cover the entire county. If awarded the permit, the county would be able to continue road improvements, among other projects. Without the permit, all road construction in the habitat came to a halt. In order to obtain the permit, the county needed a conservation plan that identified the critical habitat needed to protect each of the eight species and a means to fund it. As the clock ticked, the environmental groups were gaining momentum to sue the county and city for lack of action and protection of the species.[56] Some environmental groups and developers saw the value in showing up to meetings and working as partners. More sensational groups and individuals dressed in bird costumes to march down Congress Avenue and swoop into the city council chambers. Each group and individual played a role in gaining attention and finding solutions.

The developers against the plan quickly ran to their state representatives in hope of finding a way to continue business as usual. In 1989 the Texas legislature filed a number of bills trying to limit Austin's ability to create its own ordinances to deal with the requirements of the Endangered Species Act and zoning. According to Dede Armentrout of the National Audubon Society, "These bills are the most blatant anti-endangered species proposals to be filed anywhere in the country."[57]

All eyes were on Austin and Travis County to see how they were going to solve the endangered species and water quality issues. In 1991 US Representative Jake Pickle, stated,

> There are four major environmental efforts underway in Austin and Central Texas—the Balcones Canyonlands Conservation Plan, the Comprehensive Watershed Ordinance, The Open Space / Parks agenda and the National Parks Service—City of Austin study of the Barton Creek watershed. . . . Even though a federal law has prompted us to develop this plan, the Balcones effort is mainly a local one which requires local support. Perhaps many of you think spending millions of dollars to protect a few birds and spiders is ridiculous. Phrased that way, I admit that it sounds a little ridiculous. But think about what I have said—about how Austin's natural resources make Austin a unique place to live and work. Consider the reality that the Endangered Species Act is not going anywhere. Then consider doing something positive to help make the conservation plan a success.[58]

Barbara Carlson, a longtime activist in Travis County, was appointed as county commissioner of Precinct 2 in March 1991. Barbara took office after Bruce Todd left to become mayor of Austin. Precinct 2 had some of the most critical habitat located in it, and Bruce had been part of the team focused on finding solutions to comply with the US Fish and Wildlife requirements and allow for development outside the habitat plan.

Valarie Bristol was appointed to be commissioner of Precinct 3 in October 1991 by Judge Aleshire. The judge appointed Valarie to serve out the remainder of Pam Reed's term after Governor Ann Richards appointed Pam to the Texas Water Commission. A fourth-generation Texan, Valarie grew up in Nacogdoches before attending and graduating from the University of Texas in Austin. She took a gap year to spend time with her family in Thailand, where her father had accepted a post with the US Department of Agriculture. She then attended George Washington University in Washington, DC, to earn her master's in international affairs. While in DC, she met fellow Texan and political campaign organizer George Bristol. The couple married and had two children, Mark and me.

In 1973 the young family returned to Austin, where Valarie became active with the children's schools in the Eanes Independent School District. Valarie served on the PTA, pushed for the addition of an outdoor

Valarie Bristol (*left*) and Governor-Elect Ann Richards congratulating each other on their appointments and elections, election night, 1991. Courtesy Valarie Bristol

classroom, and insisted on planting more trees on the campus. (She still checks on her trees at Hill Country Middle School.) In 1985, as Mark and I were entering high school, she was elected to the Eanes School Board of Trustees. Then, at forty-seven years old, she applied to and was accepted into the University of Texas School of Law. She graduated after just three years in 1989—the same year I graduated from high school.

Upon joining the commissioners court in 1991, Valarie and Barbara became lifelong friends and political allies. Barbara continued to be the point person on the canyonlands planning team as her successor had been. Both commissioners inherited a complex matrix of stalled-out road projects, pressure for new development, an aging jail system, potholes, and a litany of other county issues. Valarie recruited two capable longtime friends, Jann Phenix (later Jann Brown) and Sharon Conner, to be part of her staff. Barbara also had a stellar staff that took the lead in identifying the properties within the canyonlands plan that should be acquired first.

Development had ground to a halt in Travis County for several reasons during this time. First, if a developer wanted to put in a grocery store, car wash, neighborhood, or the like in the zones identified as critical habitat, they had to work with the US Fish and Wildlife Service to obtain a 10-A permit that included an environmental assessment of the property. The process could take years and was costly. Second, the savings and loan bust threw heaps of land speculators and individuals into bankruptcy across the country, with an exceptionally high level existing in Texas. Third, the issue of water was starting to percolate to the top as the region pulled out of a three-year drought.

Like so many droughts in Texas, it was broken with the Christmas flood of 1991 that caused Lake Travis to rise by thirty-four feet in a very short period of time. The Colorado River, which flows through Austin, rose to twelve feet above flood stage. While touring the damage at Pace Bend Park in western Travis County, Valarie encountered a park employee at the fee collection booth wearing cutoff shorts and a ratty T-shirt. When she announced who she was, he responded with "Oh wow. You're like the head honcho of the whole shindig."

The tour enlightened her to just how overloved the park was and how poorly the resources, including the warbler habitat, were protected. She acknowledged it must be challenging for a ranger in cutoff shorts to be taken seriously enough to enforce the rules. The flood created an opportunity to implement some needed improvements that Park Manager Terri Siegenthaler and Ranger Dan Chapman had wanted for years. After consulting Travis County Parks Director Charles Bergh and seeing the staff's vision, Valarie championed park improvements and adoption of professional-looking uniforms, and followed the state park standard of allowing park rangers to become law enforcement officers and EMTs. Valarie also credits Charles and his team for moving away from county investment in small parks to pursuing larger parks to serve the community and protect natural resources.

One of the most perplexing issues Valarie inherited in her precinct was the Southwest Parkway Road District. Not so affectionately known as the "road-to-nowhere," it also happened to cut through prime warbler habitat and over the aquifer recharge zone. The parkway, only half finished in 1992, had been built with bonds that would have been paid for by development in that road district. Because of the economic downturn, nothing was developed and the bonds were close to being in default. The largest landowner along the road was developer and international mining company Freeport-McMoRan. Because of the road district debacle, landowners in the district, including mall developer Melvin Simon and Associates, were having trouble qualifying with their lenders to fund new projects. Two parcels along the road, called the Uplands and Sweetwater tracts, were identified as critical habitat and considered necessary to meet the obligation for the canyonlands preserve system.

Judge Aleshire; David Armbrust, attorney for Freeport-McMoRan; Jerry Turner, bond council; and Valarie flew to New York to negotiate a deal for the existing bonds so the county could issue new ones for other projects, including purchasing lands for the Canyonlands Conservation Plan. Valarie recalled, "I stood there in the banking district, among the towers of steel, concrete and glass, surrounded by a sea of humanity and thought to myself—if those little birds only knew how hard we are working to protect them."[59]

Meanwhile some landowners and developers were participating in some unsavory actions in Barbara's precinct. Hillwood Development Company, owned by the former billionaire Ross Perot, wanted to develop a large tract of land in prime bird habitat. "Although Hillwood had promised to set aside 90 acres for the warbler, the company sent out migrant workers with chainsaws to clear-cut the land of all juniper and oak. Hillwood was caught red-handed and forced to stop. But other, smaller landowners of less interest to authorities continued to clear-cut their land. . . . Several thousands of acres of habitat were lost."[60]

Each of these issues compounded the need for the city and county to prove to US Fish and Wildlife that they had to secure a funding mechanism to purchase the lands in the Balcones Canyonlands Conservation Plan to meet the obligation for protecting the species. Then the Barton Creek Greenway/Wilderness plan jumped in the mix after the National Park Service concluded its survey and made recommendations to add park and wilderness lands to the existing Barton Creek Greenbelt west of Barton Springs Pool. The estimated cost to the City of Austin to buy the land was $20 million. The estimated cost to purchase the Balcones Canyonlands was $22 million. In 1992, under the leadership of Mayor Bruce Todd and Council Member Gus Garcia, the City of Austin took a chance to place both bond proposals on the ballot. The will of the voters was positive, and they passed.

Valarie credits the swift actions of the city with clearing the way for the county to move forward with the bond package that would have fully funded the entire purchase of all the lands needed to complete the county's obligations in the Conservation Plan. However, that could not take place until the Southwest Parkway Road District was cleared up. The task force Valarie chaired worked with Freeport-McMoRan, which ultimately negotiated a buyout of the road district's debt, allowing the county to reset the taxes, so each landowner owed a proportional debt relative to the size of their property. When the debt was cleared, the landowners were freed to develop (if they had a 10-A permit) or sell their property.

With the Southwest Parkway issue solved, Judge Aleshire agreed to put a bond package on the ballot to fund the county's shared burden of the Conservation Plan. That spring Secretary of the Interior Bruce Babbitt came to Austin with Assistant Secretary Bob Armstrong, a former Texas Parks and Wildlife commissioner. Secretary Babbitt was deeply interested in seeing the Conservation Plan succeed so it could be used as a model in other

and her staff, including Jann Phenix Brown, commissioned a report that studied the economic impact of parks with Professor John Crompton at Texas A&M University. Along with public opinion surveys, the study helped her team make the case for bond packages in several counties.

At the same time, Valarie allowed her former husband and director of the Texas Coalition for Conservation, George Bristol, to share office space at the Trust for Public Land. George was gearing up to advocate for the Sporting Goods Sales Tax to be reinstated to fund state parks. He saw the value in the study and quickly adopted it to use in his briefs to the legislators. Later, George commissioned Dr. Crompton to tailor the study to just include the economic benefit of state parks.

In our 2022 interview, Valarie recalled that one of the initiatives she was deeply honored to work on was assisting the City of Wimberly to secure the iconic Blue Hole on Cypress Creek in Hays County. The property was up for sale and being considered by a developer who wanted to place 350 high-end homes around the swimming hole. Under Valarie's leadership, the Trust for Public Land worked with Mayor Steve Klepfer and Councilwomen Marilee Wood and Carolyn Nichols to secure a $1.9 million Texas Parks and Wildlife Local Parks Grant. Valarie credits the Friends of the Blue Hole for pulling together private investment to create a solid example of how public and private partnerships can preserve spaces that are valuable both ecologically and to the community.

Before her service at the Trust for Public Land, Valarie had become interested in conservation easements and how counties could use them to conserve watersheds and habitat. As county commissioner, she attended the first Texas Land Trust Council symposium at the Lady Bird Johnson Wildflower Center. Valarie joined the Texas Land Trust Council and later chaired the board of directors as the council moved from the Texas Parks and Wildlife Department to become its own nonprofit. "Chairing the council during the transition was like having a second full-time job. However, I was such a believer in conservation easements and the land trusts that it was worth it."[67]

In 2004 Valarie went to work for The Nature Conservancy as its legislative liaison and head of external affairs. Carter Smith, who, as a child, played Little League baseball with Valarie's son, Mark, was the new state director of The Nature Conservancy. He specifically asked Valarie to join the team to work with state and local governments and other land trusts, and to focus on water conservation in far West Texas.

Having earned a reputation for managing successful bond propositions, Valarie represented The Nature Conservancy to advocate for the 2005 Travis County bond package to purchase Pogue Hollow and Reimers Ranch in the western part of the county and Southeast Metro Park in the east.[68] Valarie assisted with strategizing for a successful bond proposal while Jeff Francell, also with The Nature Conservancy, negotiated the land transaction with the seller. "While at the county I learned I never wanted to run a bond campaign without first taking the pulse of voters. At the Trust for Public Land and The Nature Conservancy, I worked with the cities and counties to utilize professional polling organizations to make sure the voters were ready. And the messaging was clear."[69]

In November 2005 the voters of Travis County were ready and overwhelmingly voted in favor of the $62 million bond for parks and open space. The package also funded park improvements to a number of county parks and preserved water quality lands along Onion Creek and the Colorado River. In a press release regarding the Pogue Hollow and Reimers Ranch purchase, Charles Bergh, Parks and Natural Resources Division director for Travis County, stated, "By overwhelmingly voting 'yes' . . . the citizens of Travis County will soon own over 2,400 acres of pristine parkland. . . . This is the largest parkland acquisition in the history of Travis County."[70]

Valarie went on to assist several other counties with their bond proposals to purchase parks, water quality lands, and open space, including Jacob's Well. The well is an artesian spring that is the mouth of a deep cave system. With bond money and private investment, Hays County purchased Jacob's Well at the headwaters of Cypress Creek and the surrounding eighty-one acres to protect the springs and offer a cool source of recreation on a hot summer day. In 2023 during the drought, the springs stopped flowing for two years.

Valarie did a "self-check" in 2008 and decided she did not have the energy or desire to endure another legislative session. She informed the new state director at The Nature Conservancy, Laura Huffman, that she was ready to retire.

Valarie Bristol volunteering for Travis Audubon, circa 2013. Courtesy Mark Bristol

Retirement lasted about three months, as she quickly joined the board of directors of Travis Audubon. Not long after she joined the board, Travis County asked Audubon to sign a commitment saying it would permanently protect its 715-acre sanctuary in North Austin, which had Golden-cheeked Warblers on it, from development. The board was ready to sign a letter when Valarie asked, "Why would we do that? Let's talk to the county about a conservation easement."

Valarie's old friend from her days at the county, Clif Ladd, was also on the board. The two of them approached the county about a conservation easement. (Somewhere during this time, Valarie was named board president.) After several years of negotiation and paperwork, the county agreed to a $3.5 million easement, of which Travis Audubon put $2 million in an endowment to care for Baker Sanctuary and used another $1 million to build eco-friendly bathrooms, a sewer system, and a residence for the on-site sanctuary manager. The permanent commitment allowed the county to count the 715 acres in the Balcones Canyonlands Preserve. While president, she created a new annual fundraiser to honor a worthy conservation leader each year with the Victor Emanuel Conservation Award. The fundraiser generates annual operating funds for the organization.

Valarie's legacy to the state of Texas is almost incalculable. And she would say it is all part of a collective and collaboration. However, as is the case for many humble women in this book, it takes a leader to see each project through to the end. The Balcones Canyonlands Preserve serves as a model across the country for how other counties can protect endangered species. Yet it should be noted that the Texas legislature made sure that the funding structure could not be used for any other habitat plan in the state.

In total, Valarie helped conserve over fifty thousand acres across Texas that protect species, habitat, and drinking water. She assisted multiple counties with their campaigns to raise millions in bonds for conservation. Her legacy of loving and respecting nature lives on through her children and her grandchildren, Evelyn and Walter Bristol.

THE SEA TURTLE LADY

Ila Fox Loetscher (1904–2000)

Given her love of flying, it might seem logical that Ila Fox Loetscher would have had a special affinity for birds; however, birds were not her preference. Ila loved sea turtles and poured her relentless energy into saving the imperiled species, earning the title "Turtle Lady of South Padre."

Ila's debut into the world was interesting from the very first moment. Her mother, Sally A. Sharr Fox, went into labor on October 30, 1904, expecting to bring a beautiful baby into the world. Neither she nor her husband, Charles Fox, was anticipating twins. Ila's twin sister, Olive, arrived with a healthy weight, but the unexpected Ila weighed only four and a half pounds. Bringing twins into the world in 1904 was not an easy

Ila Fox Loetscher with a sea turtle on her dune bike, date unknown. Courtesy Sea Turtle, Inc.

task, even for the wife of a doctor. Shockingly, Sally had a second set of twins, this time two boys, three years later.

In 1922 Ila and Olive attended Central College in their hometown of Pella, Iowa. The small school did not hold Ila's interest, and she wanted to transfer to Iowa State University. Her father thought the big school and big city would only fan the flames of Ila's desire to major in drama and pursue a career in the theater—something that concerned him. Eventually, he relented, and on her first day at Iowa State, she met the handsome David L. Loetscher. In the end, it was not the lure of the theater or a dashing young man that her father had to worry about, it was Ila's newfound love of flying.[1]

Ila enrolled in flying lessons in 1928 and soon became the first woman in Iowa and Illinois to make a solo flight and obtain a pilot's license. The newspapers were captivated with the petite woman pilot dressed in her fitted khaki pants, black leather riding boots, white blouse, and leather jacket. She was an instant media sensation and a darn good pilot. Ila's family was shocked; however, her mother was soon won over and became one of her first passengers. Ila earned a living as a pilot by helping recruit other new pilots and securing potential aircraft buyers. In 1929 Amelia Earhart contacted Ila with the intent of forming an organization for women pilots. Ila agreed and joined ninety-eight other women to form the Ninety-Nines. She remained active with the exclusive club for the rest of her life.[2]

The year 1929 was a grand one for Ila. She helped establish the Ninety-Nines and became engaged to David Loetscher, who had been accepted into graduate school at the Massachusetts Institute of Technology. Ila remained in Iowa and continued to work at the airfield during their engagement but, at one point, almost called it off after oil from an airplane burned her and left boils on her face and neck. Flying was not as glamourous as it might seem.

After David graduated, the couple married in 1933 and moved to a farm in New Jersey. Ila said goodbye to her job at the airfield and settled into farm life and the demands of being the wife of a chemical engineer. Despite her love of children, an ovarian cyst and painful hysterectomy ended Ila's chances of having children. In 1954 David was diagnosed with lymphoma and passed away the following year. Ila could not imagine living on their happy farm without her beloved husband. She sold the house, packed up her red Thunderbird convertible, and drove to Pharr, Texas, in the Rio Grande Valley to live with her parents.

Her family's support and the newfound wonders of South Texas and Mexico helped her recover from the loss of her husband. But what really captured her heart was the newly developing South Padre Island. She bought a piece of property on the island and built a sleek, midcentury modern house. Soon after she moved in, her mother passed away and her father came to live with her. She immersed herself in island life and opened her home to friends and interesting people from near and far.[3]

Along the way, she met Dearl Adams, a local land developer who had become interested in the survival of Kemp's ridley sea turtles. Through his investigations into why the turtles were disappearing so rapidly from the shores of Mexico, he learned that the illegal turtle egg trade brought money to the economically disadvantaged people in the fishing villages. The turtles once numbered around forty thousand, according to the 1947 film shot by Andres Herrera that documented the Mexican beach near Rancho Nuevo where the turtles nested.[4] Adams had a vision to establish a nesting rookery on South Padre Island. With a small team of volunteers, including Ila, he spent several years bringing eggs from Mexico to the shores of South Texas.

In 1964 he legally transplanted more than a thousand eggs and counted approximately sixty that hatched and made it out to sea. In 1966 Ila volunteered to drive to Mexico to pick up a set of eggs—encased in the sand in which they were laid—and bring them back to Texas. After transplanting the eggs to the South Padre Island nesting site, the volunteers would camp on the beach to ensure the eggs were not disturbed. Ila loved the entire process. Adams asked Ila if she would keep three of the hatched turtles at her house so that they could study their growth rate and diet. The hook was set.[5]

Adams and his volunteers discontinued the transfer of eggs in 1967 after more than a thousand hatchlings had made it into the Gulf of Mexico. Now, the volunteers had to wait until the turtles matured and were ready to mate and lay eggs. Since they were not sure how long it took for a turtle to reach breeding age, they had no idea how long they would have to wait. Like most sea turtles, Kemp's ridley sea turtles return to the beach where they were hatched to lay their eggs.

Ila placed her baby turtles in children's plastic wading pools and began a lifelong process of learning about the creatures. Her showmanship reemerged as she communicated with media and communities up and down the coast to raise awareness about the turtles and the need to preserve their nesting habitat. She earned the moniker "Turtle Lady," and soon tourists were stopping by to see her turtles. Locals occasionally brought her sick or injured turtles, and she willingly and lovingly cared for them all. With support from the community, corrals were built to better house the turtles. Piers above the enclosures were included so guests could see and learn about the turtles without causing them harm or stress.[6]

In 1967 Hurricane Beulah ripped through the barrier islands and Rio Grande Valley. Ila fled inland with her dog and several turtles. Her house and the new holding pens were a total loss. She made arrangements with the aquarium in Corpus Christi to care for her turtles and returned home to rebuild. She also quickly resumed her busy schedule of speaking engagements, radio shows, and educational programs about the turtles. With help from Henry H. Hildebrand III, a renowned marine biologist and sea turtle expert who lived in Corpus Christi, Ila was able to secure permits from the US Fish and Wildlife Service and the Texas Parks and Wildlife Department to house and care for the animals.

During the 1950s and 1960s the fight to establish Padre Island National Seashore was ongoing and Ila wrote letters, took elected officials on tours, and did what she could. She was joined in that quest by two chemistry professors and environmental activists from Corpus Christi, Hans and Pat Suter. The dynamic couple were already deep in the trenches fighting the use of DDT, which was decimating the Brown Pelican populations along with other shorebirds and raptors. The Suters were the central figures in protecting the vital marine breeding waters of Cayo del Oso (also known as Oso Bay) from a development scheme that sought to drain and fill in a significant part of the bay and dredge the rest to put in a marina.

Despite overwhelming support to protect the bay, the City of Corpus Christi council members sided with the developers. Texas Land Commissioner Bob Armstrong stepped in and killed the idea. The Oso Bay issue highlighted the need for more protections for the wild and scenic spaces of the bays, estuaries, seashores, and marine wildlife.

In the 1930s David Colp, chair of the State Parks Board, had championed the idea of making the barrier island a state park. Citizens favored the establishment of a park but were opposed by developers and the oil industry. After several efforts, Senator Ralph W. Yarborough introduced a bill to establish the sixty-eight-mile-long national park along the northern reaches of Padre Island. On September 23, 1962, President John F. Kennedy signed the measure into law. It took six years to acquire the lands and settle multiple lawsuits to finally open the park, which was dedicated on April 8, 1968, by First Lady Lady Bird Johnson.

In May 1974 Dearl Adams and his family arrived for the tenth year on South Padre Island to see if the turtles had returned to the beach to nest. Adams expressed to Ila that his faith was waning that the turtles would someday arrive. But arrive they did. A female crawled to shore and was spotted by a construction worker, who quickly drove to Ila's house to fetch her. The two watched with excited tears of joy as the female, later named Alpha, deposited 117 eggs. It was the first recorded nest on the island in decades.

Ila described watching the turtle dig the hole to deposit the eggs: "There was a rhythm to it like a ballet step. In time she hollowed out a hole about twenty-four inches deep and a foot across. The digging took almost an hour. Then she started laying her eggs. Once the eggs were laid, Alpha started covering them. She used her front flippers to pass sand back to her back flippers. When the hole was filled, Alpha raised herself high on her four flippers and then let her body drop back on the sand of the nest." Ila's beautiful description of how the turtle carefully constructed the nest to secure her eggs was one of the first times the public had any insight into the life cycle of a sea turtle.[7]

For the remainder of her life, Ila, along with a team of volunteers and researchers, fought to preserve the beaches and dunes of Padre Island as a nesting rookery for the turtles. She cared for sick or injured sea turtles of all kinds that were brought to her over the years. When possible, she reintroduced them into the wild but kept others that would not survive at sea. She and her volunteers constructed corrals and pools in which the turtles could live and where visitors could come view and learn about them. Her fellow researchers, including a young Donna Shaver, studied her turtles to understand their developmental stages.

Ila Fox Loetscher with the sea turtles at her home, date unknown. Courtesy Sea Turtle, Inc.

Ila Fox Loetscher delivering a program to students, date unknown. Courtesy Sea Turtle, Inc.

When she traveled to speak or give presentations, Ila brought her turtles with her. Most of the time, they were packed into the car with her, or they were driven by her friend Evelyn Sizemore. Evelyn, also a member of the Ninety-Nines, assisted Ila in founding and sustaining Sea Turtle, Inc. In 1982 Ila even flew with her turtles to California to appear on *The Tonight Show* with Johnny Carson. She arrived with turtles dressed in costumes—like she did at most of her presentations. That was something a scientist never would have done, but Ila understood how to engage her audience in a way that evoked empathy and compassion.

On Ila's eightieth birthday, South Padre Island proclaimed an "Ila Loetscher Week" and threw a party in her honor. Newspapers across the state ran stories about this inspiring woman. "She has probably been responsible for more people becoming aware of the plight of critically endangered Kemp's ridley sea turtles than anyone in the country," said Pat Burchfield, chief herpetologist with Brownsville's Gladys Porter Zoo and the field coordinator for a joint US-Mexico Kemp's ridley turtle rescue project. "She has done more to promote the welfare of the turtles than a dozen marine biologists, and now, anywhere you go in the coun-

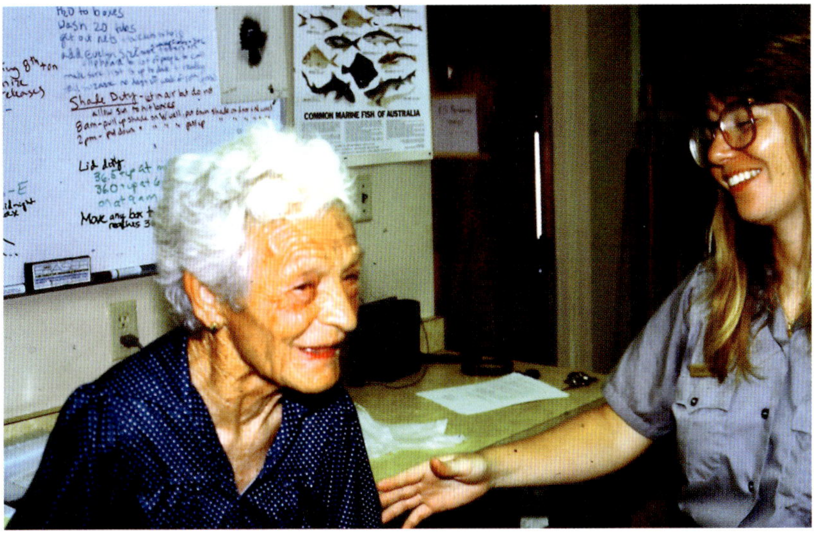

Ila Fox Loetscher (*left*) and Donna Shaver sharing a laugh, July 5, 1988. Courtesy Donna Shaver

try, South Padre Island sea turtles and Ila Loetscher are pretty much synonymous."[8]

In 1977 Ila and Evelyn founded Sea Turtle, Inc., to ensure that care of the turtles would continue after she was gone. Near the end of her life, she finally confided to her friends and volunteers that she could no longer lift the larger turtles and needed more help. In 1995 the small operation moved from her home to a new facility on the island. Ila passed away in 2000 at the age of ninety-five.[9]

Ila Fox Loetscher's legacy lives with every sea turtle nesting on the island and at Sea Turtle, Inc., which opened a grand education and conservation facility in 2018. The facility offers daily education programs and self-guided tours to thousands of visitors each year. In 2024 it completed the construction of a new sea turtle hospital. That same year they logged an astounding 10,000 volunteer hours, educated 15,165 students, rehabilitated and released 931 turtles, and protected and released 3,439 hatchlings.

In 2021 a rare cold front blasted into Texas and the Gulf of Mexico for several days. The freezing temperatures "cold stunned" thousands of turtles. An army of volunteers braved the frigid waters to rescue turtles of all sizes and species. When Sea Turtle, Inc., reached its capacity, the convention center located next door opened as a warming station. Ila would have been proud. Her legacy also lives on in others she inspired—including Donna Shaver—who now cares for the turtles that call the Gulf of Mexico home.

SCIENTIST AND AMBASSADOR FOR THE SEA TURTLES AND BEACHES OF TEXAS

Donna Shaver

Donna Shaver is the director of the Division of Sea Turtle Science and Recovery at Padre Island National Seashore. Her division is the only one of its kind in the national park system. The program focuses on studying and preserving five endangered species of sea turtles: Kemp's ridley, green, loggerhead, hawksbill, and leatherback. Her work has spanned forty years, with the majority of her career centered on the sea turtles living in the Gulf of Mexico.

While a student at Cornell University in 1980, Donna joined the Student Conservation Corps. The corps placed her at Padre Island National Seashore to help with a new program to save the sea turtles. She went on to earn her master's and later a doctorate in zoology from Texas A&M

Donna Shaver releasing the baby sea turtles safely into the Gulf of Mexico, date unknown. Courtesy Sea Turtle Science and Recovery, Padre Island National Seashore and National Park Service

From left: Ila Loetscher, unidentified woman, Donna Shaver, and Evelyn Sizemore transferring sea turtle eggs, July 1988. Courtesy Dr. Donna Shaver

University. In our 2022 interview, Donna noted that she often was the sole female in her classes and only had one female professor.[10]

The program at Padre Island National Seashore focuses on research to understand the migration, reproduction, and feeding life of sea turtles. It has also helped establish and protect nesting rookeries along Padre Island, including the one that Dearl Adams and Ila Loetscher launched. Each spring and summer, teams of volunteers patrol the beaches along Padre Island to look for returning females and help protect their nests. Donna started perhaps one of the most engaging education programs, where the public is allowed to watch, from a safe distance, as the hatchlings make their way to the surf from the nests or as they are released after being incubated. Each child and adult watches in excited hope that every turtle makes it to the surf, past the predators and fishing nets, to become an adult and return one day to resume the cycle.

Like Ila, Donna is a skilled spokesperson for sea turtles. She blends the honesty of science with empathy that connects individuals with the feeling that they can protect the turtles by using less single-use plastic or contributing to conservation efforts to care for these magnificent creatures. While Ila was certainly folksier, their core message is the same: Conservation of nesting sites is essential for the survival of the sea turtles.

Despite her demanding schedule, Donna found time to chat in February 2022. She told an inspiring story of meeting Ila Loetscher. Donna had just started working for the National Park Service and had traveled to Sea Turtle, Inc., to pick up a box of sea turtle eggs for research. She met with Ila and Evelyn Sizemore, codirector of Sea Turtle, Inc. She recounted that she was in awe of Ila and it was a wonderful memory connecting with these women, who had been the true pioneers in saving the sea turtles and building public support for their protection.[11]

In the 1990s Donna introduced a satellite tracking program. The program produced hard evidence that the shrimping industry was impeding the migration of the sea turtles and often posed a deadly barrier. She used the data to convince the State of Texas to change the laws regarding shrimp harvesting in the gulf so that they better protected the waters

A young Donna Shaver releasing sea turtles, May 1988. Courtesy Dr. Donna Shaver

around Padre Island. It was a contentious fight, but the data and science prevailed. Thanks to Donna's perseverance and focus on solutions, the new laws also required shrimping and fishing nets to have turtle extraction devices that allow turtles to escape.[12]

Even with all the technology, Donna has stated time and time again that the best tool in their quiver is the volunteers who walk the beaches each day, searching for signs of sea turtles and communicating their observations to a central hub that can collect data or mobilize resources.

In 2010 the BP Deepwater Horizon oil spill in the gulf affected all marine life, especially the sea turtles. The US Department of the Interior appointed Donna as a lead investigator for a study on the spill's impact on nesting and hatching Kemp's ridleys. As the Texas coordinator of the federal Sea Turtle Stranding and Salvage Network, Donna carefully watched the nesting cycles and educated the public on the long-term effects the spill had—and continues to have.

In 2020 Donna was nominated for a Samuel J. Heyman Service to America Medal by the Partnership for Public Service. "I believe the work that Donna Shaver has done over her career has helped save this species from extinction," said Cynthia Rubio, a supervisory biologist who works with Shaver. "Without her dedication, this species would be much worse off."[13] During her long career, she has delivered over 110 scientific presentations and authored more than one hundred publications.

But sometimes, sound science, vision, awards, and a successful program aren't enough layers of protection from those who want to eliminate programs and people who protect endangered species. Between 2016 and 2021, the National Park Service and Department of the Interior, under the first Trump administration, suffered multiple budget cuts, hiring freezes, and government shutdowns. The program Donna oversees faced a 30 percent budget cut after an ill-conceived review was released by the National Park Service. Donna did not sit back and watch her internationally and community-loved program dismantled. She contacted the Public Employees for Environmental Reasonability, which filed a lawsuit on her behalf under the federal Information Quality Act to demand that the review be retracted. For her brave actions, Donna's supervisor placed a gag order on her and tried to have her removed from her position.[14]

Donna Shaver collecting data on young sea turtles on Padre Island, date unknown. Courtesy Dr. Donna Shaver

In October 2021, a group of sea turtle volunteers and concerned citizens wrote an open letter that ran in the *Corpus Christi Caller-Times* that called for the resignation of Park Superintendent Eric Brunnemann and the restoration of grant money to the program. The letter noted that the superintendent had tried to force Donna to resign and had demanded that she not be allowed to wear the National Park Service uniform. The entire debacle was the result of a drive to dismantle various endangered species programs, even ones as successful and internationally acclaimed as the one at Padre Island National Seashore.[15] In 2023 the Sierra Club, Lone Star Chapter, reported that the superintendent's defunding of the program had resulted in the normal 25 programs per year that welcomed around 43,000 visitors to the park being cut to 5 programs per year and only reached 8,700 visitors. The reduction caused a noted drop in tourism dollars to the area.[16]

Despite the lawsuit and gag order, Donna's receipt of awards continued. The International Sea Turtle Society honored her with a Lifetime Achievement Award in 2018 for her thirty years of service. In 2020 she received the Distinguished Texas Scientist Award from the Texas Academy of Science and in 2024 she was honored with the Woman of the Year award from the Coastal Bend Women Lawyers Association. More important than awards are the numbers of sea turtles she has been able to protect over the past forty years and the countless people she has inspired to take action. In June of 2025 Donna and her volunteers located 385 Kemp's Ridley sea turtle nests along the Texas coast from Galveston to Boca Chica Beach—forty more than the year before. That's an increase of 384 nests since the turtles where reintroduced in 1974.

THE UNREASONABLE SHRIMP BOAT CAPTAIN WHO TOOK ON FORMOSA PLASTICS—AND WON

Diane Wilson

Perhaps one of the most challenging, and most important, issues to address in conservation today is pollution. Leaders in the 1970s possessed the political will and courage to take on this social and environmental injustice. However, pollution prevention and mitigation fell out of favor by the mid-1980s, when many federal and state regulatory policies were deregulated, repealed, or weakened. In 2025 another round of deregulation has swept through the state and nation. In Texas, activists must have the endurance of a marathon runner and the strength of a knight encased in a suit of armor to effect change. Diane Wilson is one of those marathon-running knights.

In 1989 Diane, a fourth-generation shrimp boat captain in Calhoun County along Lavaca Bay, became involved in pollution prevention after reading an article that identified her county as the most toxic-waste-polluted county in the country. She was familiar with the area's waterways and had made observations about how the species that lived in them were being affected. Riled by the article, she immediately knew the source—and so began her race.

Resting on the banks of Cox Creek and Lavaca Bay is the Taiwanese plastics and chemical production giant Formosa Plastics. Not far from that facility lies the now-defunct Alcoa aluminum refinery. Diane rallied other concerned citizens and launched a public awareness campaign that eventually morphed into a lawsuit. However, for people tied to the industry, it was hard to find their voice and speak their truth. In a 2021 interview with the Sierra Club, Diane reflected on that time: "Many people in the community are afraid of the powerful fossil fuel corporations that economically and politically dominate the region (even the Mayor of Port Lavaca worked for Formosa). People are intimidated when political and

complex. They are the second-largest source of micropollutants in the ocean by weight after [tire] dust. An astounding 230,000 tons of nurdles end up in oceans every year."[18]

That weight is equal to approximately 11,500 whale sharks, the largest known fish species in existence, which lives in the Gulf of Mexico. Nurdles from the Formosa plant began appearing in Cox Creek and Lavaca Bay sometime after the zero-discharge agreement was reached. Diane again rallied the citizens of the communities surrounding the bay, including a number of former Formosa employees, to form the San Antonio Bay Estuarine Waterkeeper group. Citizens, alongside Diane, waded, kayaked, and skimmed the waterways of Cox Creek, Lavaca Bay, Matagorda Bay, and San Antonio Bay to collect more than thirty million nurdles. The Waterkeepers filed a lawsuit against Formosa, claiming a violation of the Clean Water Act.

In a 2021 case study of the lawsuit, Anna Berti Suman and Sven Schade note one aspect of the unique suit: "The case—initiated by a civic group—was built mostly on citizen-collected evidence involving volunteer observations of plastic pellets, powder, and flakes in the water over a considerable time span. The contamination could not be proven through existing data held by competent authorities because the company never filed any record of pollution."[19]

In October 2019, the parties settled the lawsuit for $50 million. The settlement funds were intentionally redirected to the Matagorda Bay Mitigation Trust, not the plaintiffs (at the request of the Waterkeepers), to be distributed over a period of five years for mitigation efforts. Mitigation has included conservation initiatives, research, and education programs along the Texas Gulf Coast. In a *Texas Monthly* article, Diane stated, "We'll be involved every step of the way. If it's not resolved, we can take it to the judge. We want zero discharge. It's not just a theoretical idea."[20] Part of the district court's judgment included a ruling that if additional infractions occur and are documented, additional payments will be required to be made to the mitigation trust.[21]

The lawsuit is significant in that the plastics industry—especially in the United States—has been declaring for decades that zero discharge of effluents or solids is technologically unobtainable and economically detrimental to the industry. To put that statement into perspective, Formosa currently operates twenty production units in the US, with the largest plants found in Baton Rouge, Louisiana, and Point Comfort, Texas. The company earns annual revenues of more than $5 billion.[22]

It would appear Diane's marathon was finished; she could take a well-earned victory lap. However, in January 2021, at the age of seventy-two, she had to put on her running shoes and suit of armor again as a Houston-based oil company, Max Midstream, entered into a partnership with the Port Calhoun Authority to widen and deepen the bay's existing shipping channel. The $360 million project, which includes $120 million in bond money, is scheduled to be completed sometime in 2023, pending permit approval from the US Army Corps of Engineers (the agency that oversees dredging projects in coastal waters nationwide). The dredging will agitate decades' worth of mercury that has settled into the bay's mud and sand from the now-closed Alcoa aluminum refinery and other industries. Diane and a large group of activists opposed the port's application for the dredging permit and called for more consideration of the potential environmental damage the dredging could unleash. As of this writing, the permit remains under corps review.

In an article in the *Texas Observer*, Diane pointed out that Lavaca Bay is one of forty-nine Superfund sites in the US considered the most vulnerable due to increasing hurricanes, flooding, and storm surges. She predicted, "What the Corps, Max Midstream, and the Port Calhoun Authority are proposing by dredging the Matagorda Bay shipping channel is a doomsday scenario: a future where the bay won't be healed, the fisheries will die, and our fishing communities will become ghost towns."[23]

"For the Corps to be able to say whether the proposed project poses a threat to wildlife within the bay and potentially to human health, you've got to collect the data," explained Jessica Dutton, a biologist at Texas State University, who prepared a report supporting a call for additional review from two of the groups opposing the project, Earthjustice and the Environmental Integrity Project. "At the moment, we don't have that data."[24]

The study that eventually provided the necessary research enabling stakeholders to understand how the dredging would affect the waterways and wetlands south of the dredge site was conducted by the Harte

Institute of Research and funded by the Matagorda Bay Mitigation Trust. The proposed deeper channel is not far from the soon-to-be state park Powderhorn Ranch, which is prime habitat for the endangered Whooping Crane. Every wild Whooping Crane—around five hundred birds—winters along the Gulf Coast in Texas and Louisiana, with the majority residing in the San Antonio and Matagorda Bays. The cranes' primary source of protein is the blue crab that lives in the mud of the bays' shallow estuaries, which is where most of the toxic chemicals have settled. When undisturbed, their threat is minimized. The dredging would agitate the sediment and redistribute the toxins into the ecosystem for years.

In 2021 Diane set up a camp with a fishing boat and black coffin along the roadside to the bay and commenced a thirty-six-day hunger strike to protest the bay's dredging; for her actions, she was arrested—again.

Diane Wilson's conservation legacy is still manifesting as the Matagorda Bay Mitigation Trust invests in the community. Certainly, the study considering the impact of disturbed mercury is part of that legacy. In 2005 a documentary about her plight called *Texas Gold* was produced. The film won several awards, including Best Documentary at the New York City Short Film Festival.

In 2006 Diane published a book titled *An Unreasonable Woman: A True Story of Shrimpers, Politicos, Polluters, and the Fight for Seadrift, Texas*. In 2011 she authored *Diary of an Eco-Outlaw: An Unreasonable Woman Breaks the Law for Mother Earth*. Diane's activism has spread globally over the years as she has focused on environmental justice and antiwar strategies.

In 2013, just after the Union Carbide plant near her home in Seadrift exploded, Diane was asked to travel to India to testify against Union Carbide and Formosa in that country. Ironically, the Union Carbide plant outside Seadrift had just been voted the "safest plant in Texas" by the Texas Chemical Council. Diane stated in an interview with the Texas Legacy Project that Union Carbide had had additional leaks and often exceeded the legal amount of toxic chemicals that could be dumped into the bay.[25]

Diane is not afraid to cover herself with fake oil to prove a point to the big oil companies or testify before Congress against the corporations that claim they want personhood under the law but seek special rights when it comes to pollution. The case she and the Waterkeepers won will set the precedent for how corporations settle lawsuits in Texas and across the county regarding source-point pollution. Among her long list of awards and accolades, Diane was honored with the Goldman Environmental Prize on Earth Day 2023. The award honors the achievements of grassroots leaders and activists around the world.

A FORCE OF NATURE FOR NATURE

Terese (Terry) Tarlton Hershey (1923–2017)

Terry Tarlton was born in Fort Worth to John Tarlton, a well-known lawyer, and Elizabeth Wright Tarlton. Terry often credits her parents for instilling a love of the outdoors. After graduating from the University of Texas with a degree in philosophy, she married lawyer and US Naval Reserve Officer Thomas Hart Law. The couple returned to Fort Worth after World War II, where she followed in her mother's footsteps by volunteering for civic clubs. Terry also co-owned several businesses, including the *Westside Post* and the Wonderful Things Art Gallery, which she opened with partner Electra Carlin. In 1956 Terry and Thomas divorced.

Terry moved to Houston in 1958 after she met and married Jacob "Jake" Wilbur Hershey, president of the Commercial Transport Corporation, one of the world's largest shipping corporations. Jake was an avid sailor and the couple spent almost ten years sailing, racing yachts, and exploring faraway places such as Russia and Greece. Upon returning to Houston, Terry volunteered with the Junior League and garden clubs. However, it took a bulldozer and a bayou to shove her activist's soul out into the open.

The Harris County Flood Control District decided the best way to reduce flooding along Buffalo Bayou was to bulldoze down all the trees and channelize the circuitous waterway. As with many flood-control projects of the 1960s, the public was not engaged in the process or decision-making for how their community would be affected. Enraged citizens from the surrounding neighborhoods formed the Buffalo Bayou Preservation Association; Terry joined the group and soon took on a leadership role.

Terry began by calling on her friends and bayou neighbors, such as oilman George Mitchell and people from other organizations, including the Outdoor Nature Club and local garden clubs. Although she had resided in Houston for a number of years, she was still considered something of an outsider due to her time away from the community. Terry relied on known community presences like Sarah Emmott to spread the

Mrs. J. W. Hershey (formerly Mrs. Terese Tarlton Law), 1959. Fort Worth Star-Telegram Collection, Special Collections, The University of Texas at Arlington Libraries

word of the Save Buffalo Bayou campaign. According to a 2013 *Houston History Magazine* article, "[Sarah] Emmott was the daughter-in-law of Catherine Emmott who had waged the campaign to designate and preserve Memorial Park in the 1920s."[1]

Sarah, along with her husband, had championed the campaign to save Texas beaches from oil drilling and other environmentally destructive actions. Representative Bob Eckhardt took up the cause and, in 1959, the Texas Open Beach Law was passed. The bill protected the beaches from drilling and industry; however, it did not close them to being driven on or limit camping. Always receptive to learning, Terry engaged Sarah to help craft a successful campaign to save the bayou. Because Terry was unmoored from the social structure of Houston and was part of the economically advantaged, she felt at ease asking for assistance from all levels of society and elected officials.

The Buffalo Bayou Preservation Association was relentless in their messaging and forced the Harris County Commissioners Court into action. By 1966, it put a stop to the channelization pending further review. At the same time, Terry met George H. Bush, the newly elected Texas

Terry Hershey and George H. Bush, date unknown. Courtesy Jacob and Terese Hershey Foundation

representative. "Bush became interested in the campaign and invited Hershey to testify before the Appropriations Sub-Committee in April 1967. Successful in their appeal to Congress to order a re-study and have the US Army Corps of Engineers cease working on Upper Buffalo Bayou, Bush and Hershey were less effective in convincing local flood control authorities to reconsider destructive flood control technologies."[2]

Both Representative Bush and Terry realized the issue was much bigger than just the destruction and channelizing of Buffalo Bayou. The US Army Corps of Engineers had long since circumvented environmental impact studies and ignored the need for public comment. In 1969 businessman George Mitchell, now president of the Buffalo Bayou Preservation Association, and Terry broadened the scope of the association to include all of the watersheds in Harris County. They also changed the name to the Bayou Preservation Association. From 1966 to 1971, the association continued to educate elected officials and the public on better methods of flood management and the value of the bayous as part of that system. They also demanded that the public be part of the process.

In 1971 their efforts were successful when the federal government passed the National Environmental Policy Act, which required the US Army Corps of Engineers to include an environmental impact study and scheduled public hearings when planning major projects such as dams and other flood-control methods. Feeling the pressure of the local citizen action groups, and the passage of the new law, the corps canceled the rest of the Buffalo Bayou project. It was a pivotal moment in the history of Houston, which previously had not been successful in conservation or embracing zoning codes to protect watersheds, heritage trees, or flood mitigation.

Andrew Sansom, a friend and colleague of Terry's, reflected, "For the rest of her life, Terry credited her success in saving the Buffalo Bayou to the 'two Georges.' Thereafter, the former President referred to Terry as 'a force of nature who was green before green was cool.'"[3]

In 1970 Hershey and sixteen other women organized Citizens Who Care, later renamed the Citizens' Environmental Coalition. The coalition wanted to share best practices for creating successful environmental campaigns, notify the public and each other about issues and events, and leverage their collective power. Terry cochaired the subcommittee that reviewed national, state, and local policies that affected the environment in the Houston-Galveston region. The committee then made recommendations on how to influence the adoption of the new policies or make changes to out-of-date ones. But the ideas were before their time, and not well received in Houston. However, it was a time of great learning for Terry and put her in the position of a go-to person not just in Houston but in the state when it came to environmental policy.

In 1970 the National Flood Insurance Program was introduced as a method to minimize damages to properties in low-income communities that flooded frequently. The new program evaluated areas that flooded frequently to set national standards for flood insurance and make recommendations for limited development in those areas. Local, pro-growth governments and developers were adamantly against the idea of federal mandates of insurance and regulations. On the other hand the Bayou Preservation Association saw the value.

Terry was appointed to the Harris County Flood Control Citizens' Task Force to consider application of the National Flood Insurance Program. The rest of the members of the task force came from the development and construction industry—and none came from the affected low-income communities. Despite months of debate and lots of misinformation circulating in the media, the task force recommended applying for the program.

After her parents passed away in the early 1970s, Terry gave her family home in Fort Worth as a park and named it Wright-Tarlton Park in honor of them. Around the same time, she began taking on additional leadership roles when she was appointed by President Gerald Ford in 1975 to the Citizens' Advisory Committee on Environmental Quality and to the Board of Trustees of the National Recreation and Parks Association.

A national urban recreation study conducted in 1977 found Houston and Harris County deficient (104th in the nation) in park space. The study, as well as personal observation, prompted Terry and parks activist Glenda Barrett to rally together a new coalition called the Park People to advocate for more parks, open space, and protection of watersheds. The board of directors hired Ann Hamilton as the first executive director.

"The Park People's original mission was advocacy, and their goal to assist other groups in acquiring and maintaining park land," stated

Teresa Tomkins-Walsh in the 2013 article about Terry. "Over the years, The Park People organized projects such as Jesse H. Jones Park, drainage and irrigation for San Jacinto Battleground, landscaping for Project Row Houses in Third Ward, and wildflower planting programs. Another accomplishment was the Greenway Trails Map for Houston-Harris County, which shows 600 miles of actual and proposed greenway trails, parks greater than five acres in area, and access points for trails and parks."[4]

Terry alighted on the state stage in 1991 when Governor Ann Richards appointed her to the Texas Parks and Wildlife Commission. She was only the second woman to be appointed to the commission. At the first hearing she attended, she boldly made a statement that shocked the other commissioners and the press. After listening to an endless litany of agenda items on hunting regulations and hunting programs, she commented, "All this talk makes me want to throw up."[5] And with that, she let everyone know where she stood on that topic.

Terry's topic was parks, and she was ready to dig in and find ways to improve the chronically underfunded state park system. Terry pushed for the agency to evolve and modernize to meet the interests and demands of the public. At the time when the state was looking for ways to trim the budget, parks were always first on the chopping block. In 1992 the agency put forth a plan to close eleven state parks and limit operation at another eighteen. Terry called for an assessment of the agency's assets and engaged the media to call attention to the issue.[6]

Terry was the only commissioner to speak out and vote against allowing an oil lease and drilling platform at Falcon State Park along the Rio Grande. *The Monitor*, a McAllen-based newspaper, reported, "Armentrout [Dede Armentrout, regional vice president of the National Audubon Society] said 'It's hard to avoid the appearance that state parks are for sale.' Before the commission acted, Susan Petersen, chair of the Texas Committee on Natural resources said, 'It would be a shame if this were another case where the dollar talks and biology walks.'" The oil and gas company that applied for the lease, Sanchez-O'Brien, was owned by a former Texas Parks and Wildlife commissioner, Tony Sanchez.[7] Sanchez-O'Brien paid around $350,000 to locate three oil pad sites at the park and use slant drilling to access oil under the federally controlled lake.

Andrew Sansom and Terry Hershey at the Texas Land Trust Council conference in Houston, 2010. Courtesy Carolyn Vogel

Changing the "good-old-boy" norms both inside and outside the agency was not an easy path. In 1996, after another commissioner received a death threat from a disgruntled hunter over the issue of banning steel-jaw leg traps, Terry revealed, "I received about 100 different threats after my remarks at my first meeting."[8] But Terry persisted.

"Terry Hershey was the first and strongest advocate for state parks on the Texas Parks and Wildlife Commission," reflected Andrew Sansom. "She raised the profile of parks issues and helped her colleagues on the commission understand their importance. But she wasn't just a parks person. She understood the role that private landowners play in wildlife conservation and was the first person on the commission to herald the value of a little-known legal device known as a conservation easement."[9]

Indeed, Terry pushed the Texas Parks and Wildlife Department to implement a program to educate landowners and conservation organizations on how to utilize conservation easements. Her push modernized the conservation effort and gave landowners a valuable tool to conserve the natural heritage of their properties. Carolyn Vogel at the Texas Parks and Wildlife Department's private lands office was tapped to run the program, and Terry worked from outside the agency to promote the concept to fellow landowners.

As a founding board member of the Lady Bird Johnson Wildflower Center and close friend of the former First Lady, Terry requested that Carolyn put together a conference at the wildflower center to bring together landowners, land trusts, biologists, elected officials, and anyone else hoping to understand this new tool. Terry and Lady Bird Johnson both attended the conference. The formation of the land trusts and their conservation efforts are explored more in chapter 8.

After sailing for almost ten years, Terry and her husband were ready to sell their yacht and find a solid piece of ground to call their own that was outside the Houston area. In the early 1970s Terry's friend and fellow conservationist Lady Bird Johnson shared that a ranch, not far from the LBJ Ranch in Stonewall, was up for sale. The Hersheys purchased the 1,565-acre cattle ranch. In the 1990s they placed all of it under a conservation easement with the Texas Land Conservancy. Until the expansion of Enchanted Rock State Natural Area in 2025, the ranch was the largest piece of protected land, either public or private, in Gillespie County.[10]

After completing her term of service with the Texas Parks and Wildlife commission, Terry expanded her reach outside the state to influence national issues. She served as a trustee of the National Recreation and Park Association, the Trust for Public Land, the National Audubon Society, the National Association of Flood Plain Managers, and the National Recreation Foundation. She is also credited with helping start local nonprofits such as Houston Audubon, Urban Harvest, and the San Jacinto Air Conservation Committee, to name a few.[11]

Terry was recognized both nationally and locally for her service and advocacy. Some of the national awards she was honored with include the Cornelius Amory Pugsley Medal from the National Recreation Foundation, the prestigious Chevron Conservation Award, and the Frances K. Hutchison Award from the Garden Clubs of America. Houston Wilderness honored her with a Lifetime Achievement Award, and in 1991 she was inducted into the Texas Women's Hall of Fame.[12]

While environmental justice and conservation were Terry's main focus, she was also a strong supporter of Planned Parenthood and reproductive rights. She loved her dogs and sponsored many projects of the Humane Society. When Terry gathered up scraps from a meal at a fancy restaurant, it wasn't always her dogs she was thinking of; she often packed up a "doggy bag" to feed the wildlife in her yard.

Terry sitting with her dogs at her home, date unknown. Courtesy Jacob and Terese Hershey Foundation

Terry served for many years on the Audubon Texas board. In 2013 Audubon Texas Chair George Bristol and past chair Andrew Sansom were tasked with the mission of asking Terry whether she would lend her good name to an award. George recalls that when they arrived at Terry's home in Houston, she made them a home-cooked meal (which she asked them to help with) before sitting down to discuss the idea. Since the inaugural Texas Women in Conservation Terry Hershey Awards in 2015, over thirty women have been recognized and honored and millions of dollars have been raised to support the efforts of Audubon Texas, including the program for young women leaders in conservation.[13]

In addition to the women honored by the award named for her, Terry inspired an entire cast of women and men to step forward and take action for the environment and parks. She made it cool to be an activist and have the confidence to stand up for what you believe in and even challenge the often male-dominated power structures of politics and parks. Because she listened and cared, she spoke up for the silent managing partner of this earth—nature.

In 1991 the 496-acre Buffalo Bayou Park was renamed Terry Hershey Park. Additionally, the ten-mile Terry Hershey Hike and Bike Trail along Buffalo Bayou was named in her honor. The trail is enjoyed by thousands of people every year. The tranquil ribbon of green along the bayou offers a connection to nature for some of the most densely populated portions of Houston. Poetically, it connects to the park named for the man who once called Terry "a force of nature," George H. Bush.

As part of Terry's legacy, the Jacob and Terese Hershey Foundation has invested millions of dollars in conservation projects, primarily in the Houston and Central Texas areas or that are focused on statewide issues. The foundation concentrates on environmental justice; the health of the public, climate, air, water, and nature; reproductive justice; and the interdependent relationship between humans, public parks and open spaces, nature, and the environment. The foundation funded the research and interviews for this book.

THE GRANTMAKER WITH A VISION

Ann Thomas Hamilton

In our 2019 interview, Ann Hamilton described her childhood relationship with nature as "bipolar." Her mother actively recycled, composted, gardened, and loved the outdoors. On the other hand, her father helped design the Eastman Kodak Company's manufacturing plant along the Sabine River.[14]

In 1951 the family moved from Kingsport, Tennessee, where Ann was born, to Longview, Texas, after her father accepted the offer to help establish the new Eastman Kodak plant. She recalls attending company picnics held in the company park on the banks of the cooling lake of the plant. She watched white polyethylene pellets float across the placid water as they spilled out from railroad cars being loaded at the plant.

Ann described her family's connection to the plant as her "cross to bear as an environmentalist." She admitted, however, that her father's

Ann Hamilton being interviewed for the Texas Legacy Project with the Conservation History Association of Texas, 2003. Courtesy David Todd

well-paying job allowed her to pursue a college education and set her on the path to a great life. In the early 1960s, a career—or even a college degree—in the conservation field was not on Ann's radar when she attended the University of Colorado in Boulder. She earned her bachelor of science in home economics and a minor in business with the intent of pursuing a career in the fashion industry. Ann laughed as she mentioned that the university shut down the School of Home Economics the year she graduated.[15]

In 1965 she married Richard Hamilton, a biologist who had an interest in the environment. The couple had two children. By 1970 Ann and Richard were deep in the trenches fighting for the environment. Ann shared that she applied her "home economic skills" when she sewed the "green flag of ecology" for the very first Earth Day. She still has the flag. Ann also mentioned that the energy and mission of the first Earth Day electrified her and helped focus her desire to protect the environment.[16]

In 1974 she went to work for Governor Richard Lamm of Colorado, who had a strong environmental platform. While there, she championed projects such as Volunteers for Outdoor Colorado, which helped build trails and create access to nature.

Ann remained in Colorado for several years after her divorce in 1978, but she eventually returned to Texas. Through the help of a friend in Colorado, she landed a job as the executive director of the Park People in Houston. Ann drove with her two children from Colorado to Texas in her un-air-conditioned car to start her new job. She described driving the car around hot, humid Houston in a business suit as a miserably memorable experience.

As a single mother, with one child heading to college and the other in high school, Ann needed a job that paid more than the $30,000-a-year salary with no benefits that she earned at the Park People. When she applied to be the first executive director of the Houston Parks Board, one of the board members told her, "Don't bother. You'll never get the job. You don't know Houston." His comment angered Ann but did not deter her. She landed the job despite his objection.[17]

At the parks board, one of the projects she worked on was the acquisition of 754 acres of land for the purpose of a park in 1990. Nina Cullinan had left a bequest to the parks board to purchase the land. Twenty years

Nina Cullinan, Houston, Texas, date unknown. Courtesy David Todd

later, the Joseph S. and Lucie H. Cullinan Park was still not developed or maintained, and in 2010 several of the Cullinan heirs (Chaille Hutchinson, Emily Todd, and David Todd) contacted Ann to tour the property. Shortly after, they founded the Cullinan Park Conservancy (Ann was the first chairwoman) to create solutions to better care for the park. Through a series of negotiations, the conservancy convinced the City of Houston to enter into an interlocal agreement with the City of Sugar Land to maintain the park and make improvements. In 2018 Ann received the Nina Cullinan Outstanding Citizen Award, and into 2025 she continues her service with the conservancy.

In 1991 Ann was hired as a grants officer at the Houston Endowment, a job she very much wanted. In 1937 businessman Jesse H. Jones and his wife, Mary Gibbs Jones, established the Houston Endowment with a vision of investing in the community that brought them fortune.

In a paper titled "An Analysis of 18 Years at the Houston Endowment: September 3, 1991–October 1, 2009," Ann described her first couple of years at the endowment as a time of "history and education." She helped inventory the vast archives and visited twenty-six colleges and universities that had received endowment grants. As she did so, she built her knowledge and understanding of the endowment's intent and how best to invest in the community.[18]

Under Ann's guidance, the Houston Endowment grew its funding of conservation and environmental projects. Ann understood that having healthy, thriving citizens meant having a healthy environment. Pollution, lack of parks, flooding, and other environmental issues disproportionately affect low- and moderate-income communities. Those issues were further compounded by poor or nonexistent policies to protect the affected communities and the surrounding environment. From 1992 to 1997, the endowment invested $6 million in the environment, ranking it third in the endowment's focus areas of education, arts, environment, human services, community enhancement, and health.

From 2004 to 2009, the environment far outranked the other investment areas in the endowment's portfolio, with $49 million going to that category. During Ann's eighteen years with the endowment, she directed more than $85 million to projects that supported conservation in Texas. In total, she approved more than one thousand grants for a total of $200 million invested in the full portfolio she managed. While the primary focus was the Houston region, the funds could be awarded to statewide projects as well.

In 2004 George Bristol, executive director of the new nonprofit Texas Coalition for Conservation, sat down with Ann to reveal a plan to reinstate the Sporting Goods Sales Tax to fully dedicate it to funding state parks—as was originally intended. Ann saw the value, and the endowment awarded the Texas Coalition for Conservation a three-year grant to commission a study—the *Economic Contributions of Texas State Parks*, by Professor John Crompton at Texas A&M University. The funding also went toward conducting polls to determine whether voters favored the concept. They did. In 2005 the first attempt at Sporting Goods Sales Tax

Ann on the Ann Hamilton Trail at the Indiangrass Preserve with the Coastal Prairie Conservancy, 2018. Courtesy Ann Hamilton

legislation had good support but failed to pass because of a last-minute fight over funding for education. It failed again in 2007 just before the country slipped into a recession.

Despite having bills pass the House and Senate in 2017, it wasn't until 2019 that voters got to vote on a constitutional amendment dedicating sporting goods sales tax revenue to Texas State Parks and Historic Sites. The amendment overwhelmingly passed with 88 percent support from voters.[19]

Ann and fellow environmental leader David Todd assembled a group of Texas grant makers to form an affiliation—the Texas Environmental Grantmakers Group. The idea was to focus funding on environmental issues around the state. In 1998 the national Environmental Grantmakers Association met in Houston as a result of the affiliation. The heightened grant-making focus culminated in a project that Ann later shared was one of which she was most proud—the Texas Living Waters Project.

Water policy in Texas has been complicated, to say the least, since the first settlers built dams on the creeks running through their properties, thus limiting the flow of water to their neighbors downstream. Add thirty million more citizens into the mix and it gets even more complicated.

In 1997 the Texas legislature passed Senate Bill 1 to create a fifty-year water plan. While the plan sounded good on paper, the oversight board was filled with individuals from construction firms, river authorities, and industry consultants. The early plan met the needs of humans but did not take into consideration the needs of the ecosystems dependent on rivers, creeks, lakes, and bays.[20]

At the Texas Environmental Grantmakers Group's annual meeting in San Antonio, members of the National Wildlife Federation laid out the issues surrounding water and environmental flows. Ann and the grant makers requested that the federation craft an action plan. The resulting advocacy plan formed a coalition of the National Wildlife Federation, Lone Star Chapter of the Sierra Club, Texas Center for Policy Studies, and Environmental Defense Fund. Ann credits David Todd, director of the Magnolia Charitable Trust, with being the first funder to step forward to support the creation of the strategic plan and coalition. The Houston Endowment and other funders followed suit. The Texas Living Waters Project took on the task of changing water policy in Texas and has slowly been turning that tide.

One of the last projects Ann had the opportunity to support while at the Houston Endowment was a grant for $3 million to The Conservation Fund for the Middle Neches River project to acquire thirty-three thousand acres to connect the Davy Crocket and Angelina National Forests.

Ann "dressed to hike" at Hot Springs National Park, Arkansas, 2021. Courtesy Ann Hamilton

The board funded this up-river project; however, it was one of the last they supported outside the eight-county focus area around Houston.[21]

Early on, Ann decided that the greatest good she could do for nature was to guide investment and policy. She was able to parlay that into a career. As a volunteer, she served for a time on the board of directors of the Jacob and Terese Hershey Foundation. Ann and Terry Hershey were close friends for many years before Ann joined the board.[22]

After retiring from the Houston Endowment in 2009, Ann turned her attention to one of her other great loves—birds. She served on the board of the International Crane Foundation from 2010 to 2019 and established the Texas Crane Council to raise awareness, advocate, and fund projects to protect habitat for the Whooping and Sandhill Cranes.[23]

When asked what advice she would give women starting their careers in conservation, Ann replied, "Don't give up or get discouraged. If you get discouraged, that's OK. Go home, have a cry and a laugh, then get up the next day and get back at it. A sense of humor is very important. You have to be able to laugh at yourself. Humor has helped me get through it all." When I asked Ann what barriers she had to overcome to achieve her goals, she laughed for quite some time before stating, "You try being an environmentalist and a woman in Houston!"[24]

Ann had the vision, determination, and passion to say nature is worthy of investment. Through her savvy skills, she influenced others to do the same. While the Houston Endowment has pulled back from conservation in recent years, the state, and especially the city of Houston, benefited greatly from Ann's influence. Her work with the Texas Living Waters Project has helped highlight and underscore the need for all Texans to understand the value of water and take action to protect the life-giving aquifers, rivers, streams, and lakes.

INTENTIONALLY ENGAGING DIVERSITY

Darlene Lewis

Darlene Lewis grew up in Waco, Texas, where she attended a segregated school until entering high school. Going to college was not on her radar until her mother and a close friend talked her into applying to North Texas State University (now the University of North Texas). When asked whether playing outdoors was something she did as a child, she laughed and said, "I loved being outside. When I was outside, I could just play. Inside, there was always someone who asked you to do some chore or something."[25]

While at the university, Darlene decided she wanted to pursue a career in acting. Her friend suggested that she try radio because of her distinct voice. She agreed and interviewed for a job at the local radio station. She was hired on the spot since the station's news reporter had unexpectedly quit that same day. From 1977 to 1979, she delivered the news and learned the ins and outs of producing a radio show.

In 1979, she accepted a job with a radio station in Kalamazoo, Michigan. Her mother and friends were worried about her driving across the country alone, so her mother bought her a CB radio so she could call for help if anything happened. She packed her 1974 Mustang and embarked on what she hoped would be a lifetime adventure. The cold weather did not agree with her, so she returned to Texas in 1980 to work at K98-FM in Austin with DJ Dave Jarrot. In 1985 she moved to KOKE-FM to partner with Bob Cole for the morning drive show.[26]

She remained with KOKE-FM until 1991 when she was laid off during an economic downturn. The station did not offer benefits or retirement, which made Darlene determined to find a job that did. She was starting to feel nervous about landing something when Lydia Saldaña, a friend and the new director of the Communications Division at the Texas Parks and Wildlife Department, called her about an idea for a radio show at the agency. The two, along with a few other creative folks, put their heads

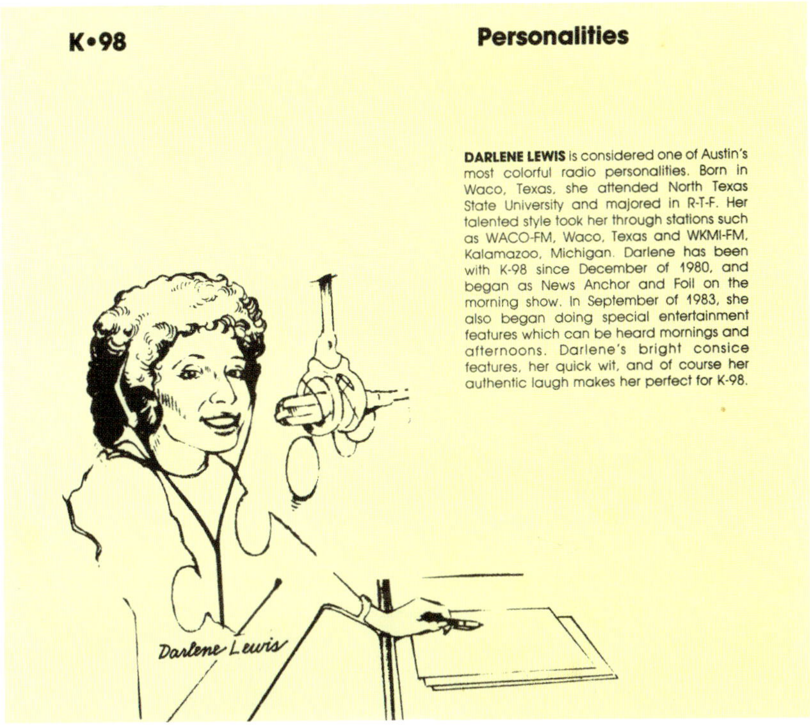

K•98

Personalities

DARLENE LEWIS is considered one of Austin's most colorful radio personalities. Born in Waco, Texas, she attended North Texas State University and majored in R-T-F. Her talented style took her through stations such as WACO-FM, Waco, Texas and WKMI-FM, Kalamazoo, Michigan. Darlene has been with K-98 since December of 1980, and began as News Anchor and Foil on the morning show. In September of 1983, she also began doing special entertainment features which can be heard mornings and afternoons. Darlene's bright consice features, her quick wit, and of course her authentic laugh makes her perfect for K-98.

Darlene Lewis

K98-FM introducing Darlene Lewis, 1980, publication unknown. Courtesy Darlene Lewis

together, and the idea was transformed into reality. Darlene was hired to be the first producer of the agency's first radio show.

Darlene shared, "I tried to make it for people like me who didn't know anything about the Texas Parks and Wildlife Department or getting outdoors. We featured a variety of people from each department and closed every show by highlighting a state park. I loved that time. I got to travel to the state parks and learn all sorts of cool things. I wanted to take listeners on the journey with me. If I was doing something new to me, I wanted them to feel like it was okay to try something new, too."[27]

Darlene also shared how much she enjoyed learning from the agency's biologists. She lamented, "No one told me I could have had a job working with bugs or animals." From that point on, she wanted to tell young people about how they could have careers in the outdoor recreation industry. "I didn't just want them to have experiences in the outdoors, I wanted them to see themselves working and volunteering there, too."[28]

In the mid-1990s, a discussion about how to engage more African Americans, Hispanics, and Asians with the outdoors was percolating to the upper reaches of state government. The Texas legislature set aside $250,000 and tasked the Texas Parks and Wildlife Department with creating a program to connect more people of color with outdoor recreation opportunities. Darlene said it was like a light bulb going off in her head. She approached Lydia Saldaña and Texas Parks and Wildlife Executive Director Andrew Sansom about an idea for a program of grants that would enable communities and tax-exempt organizations to introduce kids to hunting, fishing, camping, or other outdoor pursuits. The Community Outdoor Outreach Program (CO-OP) was born.

Darlene was tasked with developing the delivery system and structure of how the grants would be administered and awarded. For those first

Darlene Lewis holding a rehabilitated Golden Eagle before conducting the radio show for the Texas Parks and Wildlife Department in Austin, date unknown. Courtesy Darlene Lewis

few years, she did double duty, holding two demanding jobs at the agency: radio producer and CO-OP manager.

She reached out to Glen Miller in Houston and several other African American and Hispanic leaders who were offered initial grants. She asked them what they needed to sustain their programs. She also asked them to assist in setting the criteria for the grants. Collectively, they decided to focus on kids who fell into one or more of six categories: minority, low income, at risk, female, rural, and inner-urban. The funds could be used in programs that would connect kids with the outdoors, engage them in conservation education, or offer them recreation opportunities at a state park. The cohort of leaders expressed that they needed support paying for staff, transportation, food and training for volunteers and kids, and equipment. All of those components subsequently formed the core of the grant program.

After the first couple of years, Darlene saw another need. Smaller organizations were not applying because they did not have a grant writer on staff or access to one. So she developed a grant-writing training program to assist smaller organizations, especially in rural communities. Darlene shared that the training program was one of the ideas of which she was most proud, as it helped level the playing field and reach communities that otherwise would not have had the ability to fund their programs.[29]

In 1998 CO-OP funded seventeen organizations or schools to promote outdoor learning and exploration. *The Facts* newspaper in Clute, Texas, reported, "'The CO-OP program is the flagship of our commitment to children of all races and backgrounds, and our effort to introduce more people to outdoor recreation experiences,' said Andrew Sansom, TPWD [Texas Parks and Wildlife Department] executive director. 'If we are to ensure a healthy future for our state's natural and cultural resources, it is essential that we build strong support throughout the next generation of Texans, particularly those who have traditionally not been active outdoors.'" Programs receiving the grants in 1998 engaged approximately twenty-nine thousand people.[30]

Quantity was never Darlene's goal. She wanted the organizations to provide quality time in nature that kids could replicate with their families or continue as they grew into adults. "I was always talking to the

Darlene Lewis with Buffalo Soldiers on the Huff Wagon Train near Big Bend State Park Ranch, 2005. Photo by Earl Norttingham. Courtesy Texas Parks and Wildlife Department

organizations, trying to get them to stop focusing on how many kids or how many events they could include. Or how much stuff they could learn. I wanted them to focus on fun, safety, and being social with friends just as much as the lessons and activities. If it wasn't fun, the kids weren't going to do it again."[31]

Things were not always easy. She often had to educate people that "poor kids" did not necessarily equal "Black kids" (their words). At-risk kids did not automatically equal Hispanic kids. And, yes, girls did like to hunt and fish when given the opportunity. Language mattered, and Darlene called on the organizations to be more thoughtful in how they communicated to the families and the public about their programs.

Things were not always politically or financially easy either. From 1996 to 2009, CO-OP's budget grew from $250,000 to $750,000. Then the 2009 recession hit, and Governor Rick Perry cut the budgets of most state agencies, including the Texas Parks and Wildlife Department. CO-OP

Manor Middle School youth learning how to geocache while camping for the first time at Buescher State Park, fall 2012. The Camp Fire Central Texas program utilized a CO-OP grant to take kids camping and learn outdoor skills. Courtesy Jennifer Bristol

went from having $750,000 in grants to zero. For two years, the program was in limbo and almost disappeared for good.

In 2011, after considerable effort from Darlene and the organizations that participated in the grants, CO-OP was reinstated. During the dark time, Darlene listened to community leaders to consider how to make the program even more successful and easier to follow. One of the system changes she institutionalized was a requirement that all organizations interested in applying for the grants must take a training session to understand how the grants functioned, what the filing requirements were, and how to communicate their message to truly reach underserved populations.

After twenty years of championing the program, Darlene retired in 2016. She said she had found a capable replacement in Cappy Smith, who had assisted the grant program for several years after serving as a conservation education manager and park ranger in the department.

Darlene's legacy is hard to gauge in one respect. There really is no way to know how many people she inspired via the radio show to find out why

Darlene Lewis with Texas Parks and Wildlife Department Executive Director Carter Smith at her retirement party, 2016. Courtesy Darlene Lewis

life is better outside. However, we do know that from the inception of CO-OP to 2021, the agency funded 799 grants in 186 communities for a total investment of $24,195,937. The majority, 64 percent, of those funds have gone to nonprofits, which not only is an investment in the kids in the programs but is an economic infusion in the communities they serve. The rest goes to public schools. Thanks to the Sporting Goods Sales Tax, the program received an additional windfall of funding after that revenue stream grew at an unexpected rate. The cause for the growth was the COVID-19 pandemic, during which Texans turned to the outdoors to stay healthy both mentally and physically and purchased an array of outdoor gear to do so.

Today, Carley Blankenship carefully manages the program that Darlene started to ensure that children and families from all walks of life have safe and equitable access to outdoor opportunities.[32]

PHILANTHROPY AND FAITH

Martha (Marty) Leonard

In our 2019 interview, Marty shared that, as a child, she spent a lot of time outside playing various sports but did not have a strong attachment to nature. Sports played a central role throughout her life, and in the sixth grade, she was the only girl on the football team. "A fan yelled from the stands, 'Hit her hard!' And they did," Marty recollected, laughing.[33]

Eventually, she gravitated toward golf, which was a less physically aggressive pursuit. That is not to say that Marty did not pursue golf aggressively. She was consistently one of the top amateur golfers in the state and, in 2011, was inducted into the Texas Golf Hall of Fame.

Marty grew up in and around Fort Worth. Her father was an innovative businessman who owned and operated Leonard's Department Store with his brother from 1918 to 1967. The store offered everything from loaves of bread to fur coats to windmills. John Marvin Leonard (1895–1970) was also invested in real estate, was a rancher, and developed the Colonial Golf Course, which attracted the US Open in 1941. Marty inherited her parents' philanthropic values and has been in service to the Fort Worth community for more than sixty years.

Marty graduated from Southern Methodist University in Dallas with a degree in economics. Receiving her father's good business sense, she opened Leonard Golf Links Practice and Teaching Facility and Margolf Corporation. The intent of the practice center was to inspire a love of golf in people from all walks of life, not just members of a country club.

In an interview, Marty shared that her former husband enjoyed hunting, and she often joined him. Spending long hours in the deer blind, she eventually became more interested in birds than bagging a trophy buck. She revealed that her awakening to the wonders of nature came when she was on a five-day backpacking trip in Colorado with friends. She wasn't just noticing nature, she began relating to it in a more spiritual way. Marty describes herself as a woman of deep and abiding faith. That faith drives

Marty Leonard with her dog, 2016. Courtesy Marty Leonard

her to be an exceptional steward of all of God's beings. "I love every creature. Every leaf on every tree. I hope I can inspire others to feel the same."[34]

Marty has served on the board of directors of Lena Pope Home for more than sixty years. Founded in 1930, Lena Pope has evolved from a home for children in need to a state-of-the-art center that focuses on early and whole-child development for at-risk youth. In addition to education services, the organization offers counseling and behavioral support, substance abuse services, and skill-building.

In gratitude for Marty's leadership, the Lena Pope board of directors and friends presented Marty with a concept drawing of the Marty Leonard Community Chapel for her fiftieth birthday. The chapel created a space to focus on children's and families' spiritual health and wellness. The stunning chapel, designed by E. Fay Jones (1920–2004), capitalizes on elements of Frank Lloyd Wright's style that synthesize a connection with nature, faith, and design. The award-winning chapel was designed to be an "instrument for nature's music as light and shadows move and change throughout the day," a place where visitors will "think their loftiest and best thoughts."[35]

In 2014, the early childhood center wanted to improve its outdoor spaces, and Marty advocated for a more natural space, urging donors to support her vision. Partnering with Nature Explore, the center added the Marty Leonard Nature Explore Playground, one of the largest outdoor learning environments in the country at the time. The outdoor classroom and natural play area reflect Marty's desire to connect children during their early development stages with nature in the hope that they will become tomorrow's stewards.[36]

Over the years, Marty developed a passion for bird-watching, both in Texas and on out-of-the-country exploration tours with famed birder and naturalist Victor Emanuel. While on a birding excursion with Audubon in 2001, Marty was introduced to a unique place outside Cedar Hill, Texas, where she spotted a Golden-cheeked Warbler for the first time. The owner of the property and guide of the bird walk, David Hurt, had decided the dogwood-lined canyons and rare trout lilies, along with the endangered Golden-cheeked Warbler's habitat, were too special to develop and wanted to conserve them as a park.

David and his wife, Kim, recruited a few other volunteers to begin the process of donating the land and working with Audubon Texas to purchase additional surrounding properties with good habitat. The gamble was slow going until Marty arrived and saw the potential. She gave the first $20,000 gift. Her early adoption opened the doors for other donors and supporters to jump on board, until a total of $8 million was raised to complete the land acquisitions and build the facilities.

Marty at the Marty Leonard Community Chapel, 2010. Courtesy Marty Leonard

In the world of fundraising, having an early adopter who also has sway with other donors is critical. During the fundraising process, Marty helped make connections and leaned in where she could to help funders see the value in supporting the project. The 205-acre Dogwood Canyon Audubon Center opened in 2011. For her efforts, Marty was honored by Audubon Texas with the Texas Women in Conservation Terry Hershey Award in 2016.

Marty also became involved with a second conservation project that needed funding and leadership—the Fort Worth Nature Center and Refuge. In 1963 the Fort Worth Audubon Society was beginning to notice a decline in Fort Worth's bird population and a reduction in spaces to observe them as the city grew. Under the leadership of Margaret Parker and John Wilson, Fort Worth Audubon petitioned the Fort Worth Parks and Recreation Department to set aside at least fifty acres to establish a bird sanctuary and community nature center. The city dedicated lands around Greer Island on Lake Worth.

Taking advantage of the new Land and Water Conservation Fund that was formed under President Lyndon B. Johnson, the city added three thousand acres to the park and designated funds for infrastructure. In

Margaret Parker with members of Fort Worth Audubon Society at Greer Island on Lake Worth, 1964. Courtesy Friends of the Fort Worth Nature Center

1974 the first boardwalk at Lotus Marsh was constructed, with support from the Meadowbrook Garden Club and Junior League of Fort Worth and a gift from Mrs. Harry Wallengburg. That same year, a group of park advocates formed the Friends of the Fort Worth Nature Center and Refuge. Margaret Parker was the first chair. The friends group was effective for many years in constructing trails, offering programs, and raising funds for certain projects.

In 2004 Marty saw a need at the aging nature center and formed the Fort Worth Nature Center and Refuge Conservancy to focus on large-scale capital improvements that neither the city nor the friends group could tackle. Along with the other members of the board, they raised more than $3 million to improve the park's infrastructure. The friends group board hired the first executive director, Haily Summerford, who had previously worked as the director of Dogwood Canyon Audubon Center and as assistant director of Audubon Texas. Haily was also instrumental in developing the Conservation Leaders Program for Young Women.

After the Lotus Marsh boardwalk was damaged in a flood, Marty came to the rescue with financial support and rallied fellow philanthropists to rebuild it using best practices in wildlife viewing structures. In 2021, in recognition for all her work, the boardwalk and viewing platform were renamed the Marty Leonard Lotus Marsh Boardwalk.[37]

Under Marty's leadership, the friends group and conservancy merged in 2011, with a shared vision to continue to steward and advocate for the nature center. Marty was elected board president of the combined organization from 2014 to 2016 and continues to serve on the board as board member emeritus. Encompassing more than 3,650 acres that reflect the natural heritage of the region, the Fort Worth Nature Center and Refuge is one of the largest urban parks in the country and is certainly one of the few that manages an American bison herd.

Concerned about the exploding growth of Fort Worth and subsequent increase in water use, Marty ran for the Tarrant Regional Water District in 2006. The Tarrant Regional Water District has served ten counties and seventy cities for almost a hundred years, and it was run 100 percent by men from 1924 to 2006.[38] Marty beat thirteen opponents to become the first woman elected to the district and served for seventeen years. In a 2021 interview, Linda Christie, director of government affairs for the

Tarrant Regional Water District, credits Marty as the driver for water conservation efforts: "Because of Marty, we now count water conservation as a major water supply program."[39]

In our 2019 interview, Marty acknowledged the "catch-22" of the massive reservoir systems that supply Fort Worth and surrounding cities with water. For decades, the thirsty cities have relied on reservoirs as their source of water. The need to provide safe drinking water is a clear must; however, Marty also dislikes seeing land and habitat impounded for reservoirs. For that reason, she advocated for reducing demand for water by improving the Tarrant Regional Water District's ability to engage the latest technology to accurately conduct water audits and utilize wastewater recycling and reuse practices. Marty also pushed for upgraded systems to detect leaks along the 250 miles of pipes to prevent millions of gallons of water loss each year.[40]

One of the water-saving projects Marty mentioned she is most proud to have been part of is the George W. Shannon Wetlands Water Reuse Project. The Tarrant Regional Water District website states, "A joint effort with the Texas Parks and Wildlife Department, this 2,200-acre facility near Richland-Chambers Lake was completed in 2013 and consists of a series of sedimentation ponds and wetland cells that naturally filter water diverted from the Trinity River. The Richland Creek Wildlife Management Area and George W. Shannon Wetlands Water Reuse Project divert up to 95 million gallons of water a day from the Trinity River to naturally filter treated wastewater before adding it to Richland Chambers Reservoir. . . . The water depth throughout the wetlands system is maintained at eight to twelve inches to maximize water treatment and enhance wildlife habitat. In 2018, the project was recognized as a Conservation Wrangler Program by Texan by Nature."[41]

In 2019 Marty and the board updated the Tarrant Regional Water District's Water Conservation and Drought Contingency Plan, which calls for even more robust water conservation methods. Not content with mandating practices, Marty and other board leadership pushed to create training programs to share best practices and introduce conservation methods. Selling the idea to the other water board members and the community wasn't always easy, Marty admitted, but it was worth the effort, especially as the population of Texas continues to grow.[42]

A native of Fort Worth for eighty-plus years, Marty has seen the prairie ecosystem vanish. Not content to just mourn the loss, she became involved with the Great Plains Restoration Council, which is endeavoring to preserve and restore prairie at the Fort Worth Prairie Park near Lake Benbrook. The council engages children in the Plains Youth InterAction program to help restore the two-thousand-acre prairie. The hands-on program is designed to support children experiencing trauma, grief, mental health challenges, or issues at school. The children develop self-confidence, resilience, and empathy through planting, harvesting seeds, or creating rain catchment systems.

When asked what she sees as one of the greatest conservation challenges in Texas right now, Marty responded with two thoughts: "We have to do a better job of getting kids outdoors and caring about nature. And we have to preserve nature, stop paving over everything, so the next generation can have a connection with the outdoors." She also mentioned that she is concerned about overpopulation and the demand that places on the earth.[43]

Marty Leonard (*left*) and Haily Summerford, executive director of the Friends of the Fort Worth Nature Center and Refuge, 2023. Courtesy Haily Summerford

Marty Leonard's legacy lives on through the Fort Worth Nature Center and Refuge, the Dogwood Canyon Audubon Center, and the water conservation policies she helped set in motion. Her love for and care of children—and her financial investment to provide access that connects them with nature through education, exploration, and play—have affected countless youth over the decades. Those are the tangible pieces of Marty's conservation legacy. The intangible pieces—which are harder to track—come from the many ways she has inspired others to take action, to move outside their comfort zones to donate to a cause or concept that might not otherwise have been presented to them. Her calm, comforting demeanor challenges everyone she meets to see the beauty and presence of God in the song of a wren, the brilliance of a cardinal, the reflection of light dancing across the water, or the smile of a child.

FIRST LADY OF EDUCATION AND A TEXAN BY NATURE

Laura Lane Welch Bush

Plenty has been written about former First Lady Laura Bush as she championed causes that have ranged from literacy for children to public education to worldwide women's health issues. What some people might not know about her is her dedication to conservation at the national level—and in her home state of Texas.

Laura Lane Welch was born in Midland, Texas, in 1946, the only child of Harold and Jenna Hawkins Welch. Harold was a real estate developer, and Jenna worked in the family business. Laura credits her mother with instilling a love of nature in her. In a 2020 interview with *People* magazine, Laura shared, "When I got my Girl Scout bird badge, my mother was our Girl Scout leader, and from that, she [developed] a lifelong interest in birdwatching. It was slightly irritating when I was a teenager, trying to sleep in the back of the car on a long ride, and she would wake me up to look at some certain bird. But, really, it was a wonderful heritage to get that from her."[44]

The former First Lady has reflected many times on her mother's love of nature and willingness to explore the outdoors. "For almost as far back as I can remember, on warm summer nights, Mother and I would go outside and lie down on a blanket and gaze up at the stars."[45] Her mother also gifted Laura with a love of books and reading. After high school, Laura attended Southern Methodist University in Dallas and earned a bachelor of science in teaching. She taught for several years until she returned to school to earn a master's degree in library sciences from the University of Texas–Austin. She served as a librarian in the Austin and Houston public school systems before she met and married George W. Bush in 1977. The couple welcomed twin girls into the world in 1981, Barbara and Jenna.

After George W. Bush became president of the United States in 2001, First Lady Laura was able to use her voice to nationally amplify projects and programs near to her heart, including No Child Left Behind, Ready

Former First Lady Laura Bush at Monahans Sandhills State Park, 2006. Courtesy Texas Parks and Wildlife Department

to Read, Ready to Learn, and efforts to secure the rights of women in Afghanistan, to name a few. Before that, when her husband was governor and she was the First Lady of Texas, Laura helped found the Texas Book Festival along with her friends Regan Gammon and Mary Margaret Farabee. Regan Gammon also served on the National Parks Foundation board. In 2008, in recognition of the First Lady's commitment to and love of national parks, Laura was named honorary chair of the foundation. (The National Parks Foundation was founded in 1967 by Lady Bird Johnson and Laurance Rockefeller.)

After serving two terms as president and First Lady, the Bushes returned to Dallas, where Laura soon turned her attention to building

Shield Ranch in western Travis County, June 9, 2023. Photo by Jennifer Bristol

CHAPTER EIGHT

THE LAND TRUSTS

Texans love the land. However, Texas political leaders haven't always loved public lands. Even after the establishment of the state park system, it took one hundred years to curate the current collection of parks. In 2023 that system, even with the wildlife management areas and state natural areas, is just over 1 million acres in a state that consists of 142 million acres.

Cities and counties face their own legal and budget restrictions to add meaningful amounts of parklands. However, it is the cities and counties that need the ability to meet the requirements of federal mandates, or to satiate the demands of thirty million residents who primarily reside in urban areas. For decades, public parks were the only method to conserve habitat, species, water, and soil. But it was never enough.

Conservation easements became a new method for private landowners to enter into the conservation matrix while also maintaining their properties as working lands. And with over 650 acres lost per day to development it was time for a new solution. In order to make all this

Jenny Sanders and son Nate at lookout mountain, Temple Ranch, Duval County, February 25, 2012. Courtesy of Jennifer L. Bristol

function, land trusts had to develop with specialties for selected regions or types of conservation needs. This method especially resonated in private land–centric Texas. By 2025 over 2 million acres of land have been conserved by Texas land trusts.

MOVING BEYOND THE KNOWN TO FIND INNOVATIVE SOLUTIONS

Carolyn Vogel

Carolyn Vogel is a fifth-generation Texan who hails from Gillespie County. Carolyn grew up playing with friends and family along the banks of the Pedernales River, just a skipping stone's throw from the LBJ Ranch. As the firstborn of five siblings, she was often called on to be in charge. While living in the country might seem idyllic, Carolyn admits that as she entered high school, she envied the girls in town who lived close to their friends and activities. When friends did come over, they would camp or hang out by the Pedernales River. Despite living on a peach orchard, Carolyn said, "I did not invite friends over to pick peaches. That would not have been cool."[1]

When Carolyn was twelve, her future was set on an unanticipated course by her mother, who signed up young Carolyn to be a docent at the newly opened President Lyndon B. Johnson Birthplace Foundation in Stonewall. Later, when the LBJ Ranch State Park and Historic Site opened in 1970, Carolyn played in the high school band that was part of the celebration. The connection gave her a sense that "history was alive and deeply connected to the natural world."[2]

She studied political science, history, and German her first couple of years in college while at North Texas State University. However, she became discouraged when the only career options presented to her were degrees in teaching or writing. She returned home on Christmas break and informed her parents that she wanted to transfer to Texas A&M University to pursue a degree in parks and recreation.

"Most of the people in my classes wanted to work for the national parks, but not me," she noted about her time at Texas A&M University. "The professors really pushed the national parks and forests, but I wanted to work at the Texas Parks and Wildlife Department. I encouraged the professors to bring in staff from that agency to give presentations, I made

Carolyn Vogel by the Frio River at Garner State Park, 1980s. Courtesy Carolyn Vogel

those connections, and it worked. I graduated and was hired as an interpretive ranger at McKinney Falls State Park, which opened the year I was hired [1976]."[3] That same year, Carolyn married and became Carolyn Vogel Scheffer.

Carolyn worked her way up the state parks ladder, taking her second job at Goliad State Park and Historic Site, where she became curator of Mission Espiritu Santo. From there, Carolyn was tasked with opening and operating the Landmark Inn State Historic Site in Castroville. Coming full circle, she returned to the LBJ Ranch State Park as the first woman assistant superintendent.

Carolyn was not the first woman superintendent; however, she was one of only a handful who have served in that capacity since the first women stepped into the role in the 1940s. During World War II several women broke down the barriers to having a career at the state parks. According to a 1946 article in the *Waxahachie Daily Light*, "The State Parks Board has a college president and former governor as one of its park managers (Pat Neff). There are four women managing state parks. Mrs. Ester D. Webb is boss at Caddo Lake Park; Mrs. Douthitt Alcus,

Carrie May Ferrell at Stephen F. Austin State Park, 1947. Courtesy Texas Parks and Wildlife Department

manager of Davis Mountains Park; Mrs. Ethel W. Harris of San Jose Mission, and Mrs. Carrie May Ferrell of Stephen F. Austin State Park at San Felipe."[4] Douthitt Alcus worked and managed the Davis Mountain State Park and Indian Lodge for over six years.

While at LBJ Ranch State Park, Carolyn received a call from the Texas Parks and Wildlife Department Regional Director Tom McGlathery, who suggested she apply for a regional director position in the Houston area. The position was closing at the end of that day, so Carolyn hustled to pull together her résumé and paperwork to apply for the job. She called her husband to say she had applied, and he enthusiastically replied, "Great! Let's go to La Porte."

After the interview, Carolyn waited a long time on pins and needles as the powers that be deliberated over placing a woman in a regional leadership position—something that had not been done before. Thankfully, an ally championed her and argued that she was the most qualified. "It's funny what you remember. In the interview, I was even asked if I would

Regional Director Carolyn Vogel Scheffer firing the canon at San Jacinto Battleground State Historic Site at the opening volley of the Sesquicentennial Celebration, 1986. Courtesy Carolyn Vogel

be OK driving in Houston traffic."[5] In 1985 Carolyn became the first woman regional director, almost sixty years after the formation of the state park system.

Shortly after moving and jumping feet first into the position of overseeing thirteen parks in Southeast Texas, Carolyn, along with her husband, welcomed a daughter, Kassi Scheffer, into the world. In addition to becoming a new mother, Carolyn was in the spotlight as co-operations manager of the Texas sesquicentennial birthday event at the San Jacinto Battleground. The San Jacinto Battleground and Battleship *Texas* were both located in her region. And the battleship was sinking into the mud of the Buffalo Bayou. Carolyn said she loved being part of the team selected to ride on the Battleship *Texas* as it was towed out of the mud and down to Galveston for repairs in 1989.

When asked whether she was afraid it would sink, she replied, "Of course! We had no real idea how bad the damage was because it had been in the muck for decades. We just smiled and said a few prayers and waved at the cameras. The engineers in charge prepared well, which is fortunate because the second we got into the ship channel, we started taking on water—a lot of water. But it made it."[6]

In 1991 Carolyn applied for the position of chief of park operations, which was stationed at the headquarters in Austin. In our interview, she stressed that at that time, her entire goal in life was to be director of the State Parks Division. She was hired as chief of park operations and became the first woman in that position, but all was not rosy.

"Basically, the director of the State Parks Division at the time did not want to hire me," she remembered. "I was not his first choice, but others championed me because of my years of park operations experience. He was not happy about it, so I spent the next eighteen months working six

Carolyn Vogel (in white dress) with state park staff and volunteers at San Jacinto Battleground State Park, date unknown. Courtesy Carolyn Vogel

and seven days a week to prove my worth and kept hoping he would change his view. I thought if I just worked hard enough, things would change." Understanding her position was unsustainable at that rate and in a toxic environment, she made a change and took a step down the ladder to become a state parks regional director in Kerrville in order to preserve her sanity.[7]

Carolyn says she was happy being back in the Hill Country; the only issue was that her husband and young daughter were in Buda, which made it a challenge to get home in time to pick up Kassi from school or be there for other events. Putting family first, she made another change in 1993 and went to work in the conservation program at the Texas Parks and Wildlife Department as a land acquisition specialist. It was a good fit.

Terry Hershey, a Texas Parks and Wildlife commissioner at the time, proposed an idea that would become an even better fit for Carolyn and the state. Commissioners do not often appear at staff members' offices and make requests, but Terry did. She approached Carolyn's boss, Mike Herring, and asked whether he could find out more about conservation easements. Moments after Terry left Mike's office, he hustled down to Carolyn's cubicle. She recalled, "He said, 'I need you to find out everything you can about conservation easements.' I was like, 'OK, but what are they?' So, I signed up for a conference in Tennessee and attended an eight-hour workshop presented by Professor Bill Hutton of the University of California, Hastings College of the Law. About twenty minutes into the seminar, I thought, oh boy, bringing this conservation easement methodology to Texas was going to be a challenge."[8]

In typical Carolyn fashion, she learned all she could. In the process, she realized there was a need for more training and awareness in the agency and among landowners. "At the time, landowners were coming to our office every week wanting to sell their land to the state for a park or wildlife management area. Some had great property with wonderful habitat or river access, but we didn't have the money. It was hard to turn them away with no solution."[9]

Carolyn's first training session was held in 1994 for Texas Parks and Wildlife staff. After that first session, she started to build outward, pulling together the few land trusts working in Texas, including the Valley Land Fund and The Nature Conservancy. Carolyn recalled that the conference she held in 1996 was a "bellwether" for her, as it brought together landowners, land trust volunteers and staff, lawyers, realtors, and other people who had been naysayers against the use of easements.

In our 2022 interview with Carolyn and Valarie Bristol, former president of the Texas Land Trust Council, Valarie recalled of that time, "The naysayers had all kinds of misinformation they were putting out. They told people that conservation easements were just a way for the government to take your land. That the government was going to swoop in and devalue your land and then come back with an offer that was pennies on the dollar. It was complicated, so that made easements seem mysterious and open to distrust."[10]

Carolyn dispelled the distrust and built a collective of land trust organizations to which she and others could refer landowners. Under pressure from Terry Hershey and fellow commissioner Mickey Burleson, Carolyn's boss shifted her position to focus exclusively on creating the trainings, raising awareness, and building the network. Now, the agency had a tool they could hand landowners who wanted to preserve their lands for future generations and the partners who could aid them in their goals.

Land trusts are defined as charitable conservation-focused organizations that assist landowners in meeting their long-term land use goals and protecting their legacy through a variety of tools such as land acquisition programs and conservation easements. Conservation easements restrict development rights to protect the land and water, while the willing landowner, who is in the driver's seat, continues utilizing the property as a ranch, a farm, a getaway, or simply a functional ecosystem. Land trusts do not rely on tax dollars but are instead nonprofits that can stitch together public and private partnerships. Each conservation easement is tailored for the landowner and the conservation goals of the land trust.[11]

A county or city can also enter into a conservation easement with a landowner, and those funds can come from a variety of sources, including bonds, sales tax, federal programs, and disaster mitigation dollars such as the BP Deepwater Horizon oil spill restoration funds.

One of the first land trusts in Texas was created by attorney and naturalist Ned Fritz and was called the Natural Area Preservation Association—and eventually the Texas Land Conservancy. By the mid-1990s, more such groups were emerging around the state. Some of the most effective ones focused on a specific ecoregion within a limited range of counties. For example, in 1998 the Coastal Bend Land Trust was established by the Coastal Bend Bays Foundation to "preserve and protect ecological habitats in the Coastal Bend watershed areas," said Jennifer Lorenz, the foundation's executive director. "There are no land trusts in this area. It's a creative answer to today's land conservation challenges."[12]

Those challenges were coming faster and faster. In the 1990s Texas led the nation in converting farms, ranches, prairies, forests, and wetlands into shopping centers, houses, roads, factories, and office buildings. By 2002 the US Department of Agriculture estimated that Texas was losing around 283 square miles of land every year to development. Much of that land was located in critical ecosystems and watersheds. As Carolyn noted in an *Abilene Reporter-News* article, "The important message to landowners in Texas is they have choices. I think sometimes landowners feel that their only resource is to sell out to asphalt."[13]

Still working for the Texas Parks and Wildlife Department, Carolyn became the coordinator for the Texas Land Trust Council. Each year, they held a conference to continue the trainings and expand the number of land trusts. In a 2002 article in the *Corpus Christi Caller-Times*, Carolyn stated, "Since the 1990s, the number of private land trusts in Texas has more than doubled, from 17 to 39, with more of them planned."[14]

One of Carolyn's greatest gifts was mentoring the new land trust organizations and their staff. In 2014 Carter Smith, former executive director of the Texas Parks and Wildlife Department, said this about her mentorship:

> I first had the great pleasure of meeting Carolyn when I moved back to Texas from Canada to assume a position with the Katy Prairie Conservancy (KPC), a then-fledgling land trust organization with no office, no staff, no money, and, ahem, no land actually conserved. . . . I arrived young, naive, eager, and blissfully deficient in

The Texas Land Trust Council honoring Carolyn Vogel for her years of service, 2018. *Left to right:* Kassi Scheffer-Geeslin, Carter Smith, Texas Land Trust Council Director Lori Olson, and Carolyn Vogel. Courtesy Carolyn Vogel

the ways, means, and machinations of land trusts and their work. Candidly, I had no real conception of what I was supposed to be doing. Thank God for Carolyn and her steady and endless mentoring, or else I might be back researching trophic cascades and counting moose in the furthest reaches of the boreal forest in northern Saskatchewan. With her deep ties to the lands of the Hill Country, formidable German roots, Aggie pedigree, long and successful tenure with the Texas Parks and Wildlife Department, infectious personality, maternal ways, and strength of spirit and character that was simply unassailable, Carolyn was just the leader to usher the land trust movement into Texas.[15]

With a strong, well-trained, and well-mentored coalition, the Texas Land Trust Council could begin tracking the success of the program. By 2003 land trusts had conserved 312,000 acres set aside under easements or acquired through fee simple purchase. And then, in 2005, Carolyn and the Texas Land Trust Council were shocked when the Texas Parks and Wildlife Department informed them that they could no longer be housed within the agency nor be supported by staff. The edict came mostly from the Legislative Budget Board, which was "cleaning up" some of the state agencies' methods of doing business. This often included supporting coalitions with staff time and resources. Carolyn chose to retire from state government and become the founding executive director of the Texas Land Trust Council. Following eighteen months as executive director, she determined that a break was needed, and she stepped down.

The Texas Land Trust Council had a capable board of directors, which included Valarie Bristol, Kathryn Smyth, James Jeffries, James Varnum, June Kachtik, David Bezanson, John Herron, Joy Hester, Colleen Curran Hook, Norene Hutchinson, Nan McRaven, and Todd Votteler. Despite a bumpy start, they carefully positioned the new nonprofit for success and carried forward the movement Carolyn had put in motion.

When I asked Carolyn what she would say to a young person wanting a career in conservation, she replied, "Do what you can, partnered with a sense of purpose. You don't have to change the world. But if you focus on the good you can do that day or month or year, then it starts to make a difference. If you can't get the job that you want right out of the gate, volunteer for that organization or something similar to gain experience and make connections. It's all about connections, not just to get hired, but to be impactful. I think a sense of purpose is key to a meaningful life."[16]

Carolyn added something profound about being a woman of her generation and what it felt like to be the only woman in the room sometimes. "I was born in the 1950s and schooled in the 1970s. There was a time that I convinced myself that women, including me, were second-class citizens. As an analyzer, I could not figure out any other reason why I was being treated the way I was. I must be second-class, and I accepted that I was less than. Isn't that sad?" Valarie Bristol agreed and added, "You're right. I never thought of it in those terms. It was subtle, and I accepted it as just how it was. I guess we both grew out of that, thank goodness. But it took a long time."[17]

Beyond the achievements of women like Carolyn and Valarie, their generation secured acceptance of women being in the room and being part of the conversation. They weren't always heard, but every day that they showed up was one day closer to shifting the norm.

When I asked Carolyn what she thought was the biggest conservation issue facing Texas at this time, Carolyn first replied, "Water." Then she revised the remark to say, "We have to change attitudes. People need to understand that water doesn't come from the faucet. Food doesn't come from H-E-B. I'd like to see a shift from materialism to the values that my parents modeled: a love of the land, a simpleness of spirit, a desire to leave things better for those not yet here. We have to shift people's short-sighted mindsets and help them understand that each of us is a steward of this earth."[18]

Thanks to the push from Terry Hershey and Mickey Burleson, Carolyn's legacy moved out of the boundaries of state parks to become one of the best solutions for conservation in Texas. That legacy lives on in the Texas Land Trust Council and all the land trusts that she advised and supported as they built their own capacity. Today, the Texas Land Trust Council is directed by Lori Olson and a stellar board of directors. The council tracks the conservation land inventory, reporting in 2025 that since the 1990s, more than 2 million acres of land have been secured through conservation easements or have been purchased by a land trust. That is roughly 1.2 million more acres than all the state parks in Texas combined.

A 2019 report titled *Valuing Economic Benefits of Texas Conservation Lands*, produced by the Siglo Group for the Texas Land Trust Council, states that conservation lands provide $30.9 billion in flood prevention, water supply containment, and water filtration.[19] The council continues to educate landowners; state, city, and county staff; elected officials; and trusts at its annual conference. The council provides technical support, advocacy, and guidance for interested landowners and connecting them with other landowners.[20]

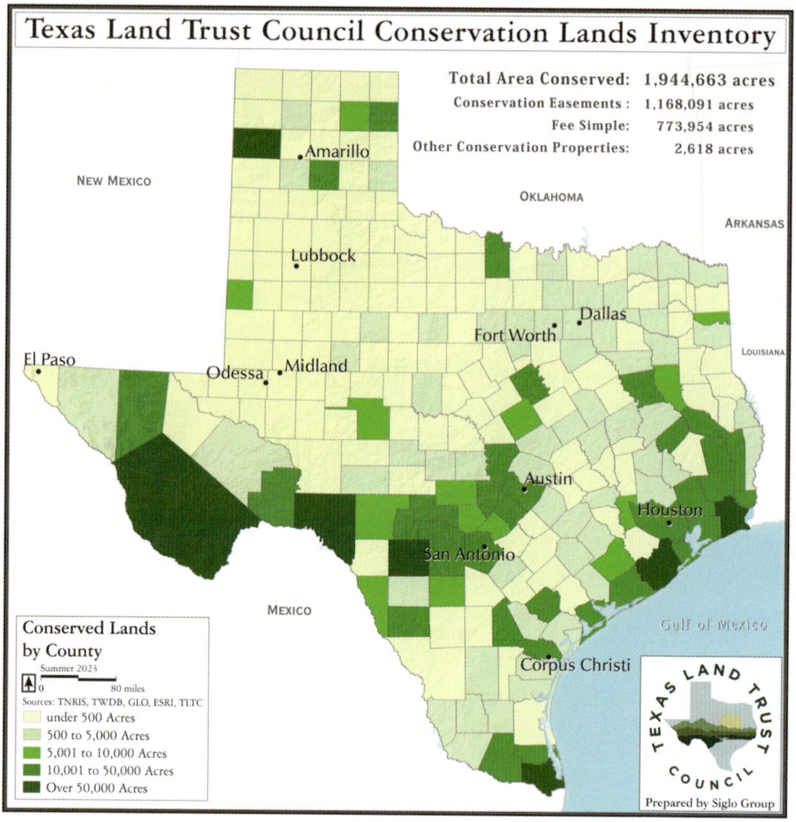

Texas Land Trust Council Conservation Lands Inventory, 2023. Courtesy Texas Land Trust Council and Siglo Group

Carolyn's legacy lives on in her daughter, Kassi Scheffer-Geeslin, who has been the director of education and outreach programs at the Texas Wildlife Association for more than fifteen years. Kassi shared, "Being a daughter of a conservationist and the daughter of Carolyn Vogel and working in the conservation field . . . meant having a first-hand portrayal of a respected woman comfortable with standing toe-to-toe with colleagues in a primarily male-dominated industry. She showed me the value of dedicating to a cause you believe you can make a difference in. She also showed me how I should work to ignite the 'conservation flame' in the next generation of Texans."[21]

Carolyn Vogel, land broker, 2021. Courtesy Carolyn Vogel

While conservation easements and land trusts have been a win for landowners wishing to conserve their lands and municipalities seeking ways to preserve their resources, the trusts also have been a space where women could excel. There were—and are—certainly many wonderful and skilled men in the field; however, the nonprofit sphere attracts many innovative, conservation-minded women who might have otherwise been shut out of traditional conservation careers.

Today, Carolyn is a real estate broker who focuses on farms and ranches, offering expertise in selling properties with conservation easements or to conservation-minded buyers. In 2023 the Texas Parks and Wildlife Department celebrated its one hundredth anniversary; however, there still has never been a female state parks director or executive director—yet.

DON'T PAVE THE PRAIRIES

Mary Anne Piacentini

Mary Anne Piacentini became a conservationist via a circuitous route. According to her hometown newspaper, the Portland, Maine, *Evening Express*, "Mary Anne Piacentini, daughter of Anna Mary (Mangino) Piacentini of 6 Fobes St., went from Harvard to Houston armed with a master's degree in city planning. But Houston wasn't a city much interested in city planning at that time. Developers, unhampered by zoning laws, were running rampant."[22]

How did a city planner end up with a lifelong career in conservation? She worked for a short time as a housing and policy planner for the City of Houston in the 1970s. From there, she went on to direct the Cultural Arts Council of Houston and then, between 1993 and 1997, she guided the Friends of Hermann Park through a massive master-planning and restoration effort to improve the 445-acre historic urban park. She progressed from being a one-woman show to having a staff of seven, overseeing development of the comprehensive master plan that identified more than $75 million needed for capital improvements in the park.

In 1992 the Katy Prairie Conservancy was founded by a collection of hunters, birders, and concerned individuals who were alarmed at the rate at which the prairie was disappearing. Prairies have traditionally acted as giant sponges that capture heavy rain loads and slow water runoff, thus reducing flood events. But flat, easy-to-clear prairies are seductive to developers looking for lands on which to plop down a bunch of houses, strip malls, warehouses, and roads. However, the Katy Prairie and pockets of wetlands are a haven for wintering ducks, geese, Sandhill Cranes, shorebirds, and other wildlife. As refineries and industry filled the land east of Houston, housing developers looked west to fulfill the suburban dream.

In 1999 Mary Anne joined the Katy Prairie Conservancy as their executive director. Keen to learn more about how land trusts and conservation easements worked, she joined the Texas Land Trust Council and met

Mary Anne Piacentini, 2015. Courtesy Mary Anne Piacentini

Carolyn Vogel. She later reflected on joining the conservancy and being thrown into the new wild west of land trusts, "I had never directed a land trust, so I was eager to learn as much as possible. Carolyn was patient and extremely generous in sharing her knowledge not only with me, but also with many others concerned about conservation in Texas."[23]

Mary Anne could see that conservation easements were going to be a game changer. She joined the Texas Land Trust Council board (and chaired it in 2004) with a goal to help raise awareness of the generational value of easements and preserves. It did not happen overnight, especially in Houston, where developers were used to getting what they wanted.

One of the dilemmas that Mary Anne faced when she began working for the conservancy was how best to protect the prairie from a road project called the Grand Parkway, a 170-mile loop planned to encircle Houston. The idea for the highway had been floating around since the 1960s. In 1984 a special session of the Texas legislature increased funding and created the Grand Parkway Association, which, according to an article in the *Galveston Daily News*, was "a state-funded nonprofit corporation that lobbies county governments to help build the project. . . . Usually, before the state transportation department undertakes a road project, it conducts exhaustive studies and public hearings to determine whether it's a good idea. But with the Grand Parkway Association, the process seems to have been reversed. The state has decided up front that the project is a good idea and that it should cultivate public support. Only after that will engineering and environmental studies be done."[24]

Under Mary Anne's leadership, the Katy Prairie Conservancy was adding acreage to its inventory of prairie as fast as funding would allow. By the mid-2000s it was becoming clear that the Grand Parkway, the third major loop around Houston, was going to be built, and it would cut through the prairie, subjecting it to even more development and fragmentation. Mary Anne stepped up her efforts to raise awareness and educate elected officials and developers about the invaluable ecosystem services the prairies provided.

The road was marketed to the public with the idea that it would have limited access and exit points because it primarily would serve as a hurricane evacuation route, and that it would provide an efficient transportation corridor for trucks to avoid traffic in Houston. Mary Anne and the board took a neutral stance regarding the road but noted that if it were eventually built, the Texas Department of Transportation should ensure that the environmental impact of the project would be limited.

The Katy Prairie Conservancy and other environmental groups recommended that mitigation be robust. Mary Anne continued her work to develop public support for appropriately scaled mitigation, while the Sierra Club sued, unsuccessfully, to stop the project altogether. In the end, engineers modified the road somewhat and shifted the layout farther to the east, away from Katy Prairie Conservancy lands. However, the other concerns that conservationists feared came to fruition: The parkway ultimately exposed the prairies both north and west of Houston to rampant growth.

To put the growth in perspective, the population of Waller County in 1990 was 23,390. In 2020 the population had grown to 56,794. However, those numbers seem small compared with the behemoth to the east. Harris County's population grew from 2.8 million in 1990 to 4.7 million in 2020. It is worth noting that the city of Katy straddles the line between Waller and Harris Counties. Additionally, the population center of Houston has shifted to around the Barker Cypress area, which is far west of downtown Houston and much closer to the prairie.

Hurricane Harvey hit in August 2017, dumping between thirty-three and forty-eight inches of rain across the Houston area. Even though the city was not in the eye of the hurricane, it received the majority of rain. The City of Houston, Harris County, and surrounding counties finally began to hear what Mary Anne and teams of other experts had been saying all along: "Don't pave the prairie."

The disaster of the hurricane finally pushed the value of prairies and upstream land conservation to the forefront for Houston. Mayor Sylvester Turner included prairies in the city's long-range resiliency plan. "Being included in the resiliency plan is a result of Mary Anne's relentless work to inform and educate everyone from Rotary Clubs to private landowners to elected officials. Really, it's an artifact of her persistence that will pay dividends in the future," noted Jaime Gonzalez, former community conservation director for the Katy Prairie Conservancy.[25]

Mary Anne extended her reach beyond the boundaries of the preserves to serve on the Harris County Flood Control District's Cypress Creek Overflow Committee. After the committee submitted their findings, the US Army Corps of Engineers began the Buffalo Bayou and Tributaries Resiliency Study. The Corps of Engineers submitted their own flood-mitigation proposal of tried and tired ideas, many from the corps' 1940s plan.

Mary Anne engaged the community and helped fund a separate study to offer alternative flood-mitigation solutions based on the findings of the committee. She and the committee proposed to work *with* nature instead of trying so hard to *control* nature to slow water flow, to retain floodwater more efficiently within existing natural infrastructure. The plan called for an investment in conservation easements that would allow aboveground vegetation and the deep roots of prairie grasses and tree groves to impede and absorb the runoff from heavy rain. That system also would provide the added ecosystem service of carbon sequestration, which fits into the Harris County and City of Houston climate plans.

Mary Anne met with various committee members in the Texas legislature to encourage them to consider the benefits of nature-based solutions and the importance of retaining natural infrastructure to help slow down, absorb, and store floodwater. She noted that these efforts would not eliminate flooding but would help mitigate its impacts. She also noted that many of the engineering solutions that had been tried for many years had not eliminated flooding either, so the plan suggests that a hybrid approach that harnesses nature and engineering together might be a more effective solution.[26]

The vision to move beyond the conservation of just the Katy Prairie to include all Houston-area coastal prairies had long been on Mary Anne and her team's radar. As Jaime Gonzalez remembered, "Mary Anne is a dynamic and creative thinker and leader who isn't threatened by other people's ideas or sharing resources that might benefit the greater good. When I worked at the Katy Prairie Conservancy, she allowed me to cofound the Coastal Prairies Partnership with conservationist and native prairies advocate Flo Hannah as a means to bring together everyone working with prairies in the greater Houston area—and beyond. Each year, we hosted a conference to learn from each other. She [Mary Anne] could see the power of the collective impact. I know she had wanted to expand for a long time. She finally was able to move the conservancy forward in 2022 to become the Coastal Prairie Conservancy to scale its scope."[27] The Coastal Prairie Partnership, which was already part of the Katy Prairie Conservancy, dissolved and became part of the new structure.

The expansion into nine counties was accompanied by a rebranding as the Coastal Prairie Conservancy and opened up opportunities to work with landowners along the marshes and coastal prairie systems, an at-risk ecosystem that is crucial to fortify the region against hurricanes and rising sea levels. Increasing the scope unlocked new funding sources, including those provided by the RESTORE (Resources and Ecosystems Sustainability, Tourist Opportunities, and Revived Economies of the Gulf Coast States) Act, a 2012 federal law that awards funding from the 2010 BP Deepwater Horizon oil spill settlement to support restoration projects, preserves, and conservation easements along the coast.[28]

Under Mary Anne's leadership, the Coastal Prairie Conservancy's overall land holdings have grown from thirteen hundred acres to more than thirty thousand acres. The organization's goal is to increase protected lands to nearly sixty thousand acres. The conservancy's lands form a matrix that serves various purposes; for example, some units are open to the public, including the Indiangrass Preserve and Ann Hamilton Trail, as well as the Matt Cook Memorial Wildlife Viewing Platform on the historic Warren Ranch. Other units are wetlands and protected bird habitat, while still others are conservation easements on working ranches. In addition, the Coastal Prairie Conservancy and Mary Anne have long placed a high value on outreach and have provided education opportunities for families and adults both at the preserves and in Houston and local school systems.

Always willing to engage in environmental-based, citizen- and community-focused coalitions, Mary Anne has served on the Houston Galveston Area Council of Government's Natural Resources Advisory Committee; Coalition for Environment, Equity, and Resilience; Waller County Strategic Planning Steering Committee; Waller County Mobility Study; Land Trust Accreditation Commission; Land Trust Alliance Energy and Climate Working Group; and Land Trust Alliance Leadership Council. She also served on the advisory board for the Texas Coalition for Conservation, which pushed for the full reinstatement of the Sporting Goods Sales Tax to fund state parks and state historic sites.

Mary Anne founded a multiyear collaboration called the Headwaters to Baywaters Initiative to protect riparian corridors. She is currently chair of the Conservation Flood Mitigation Group and in 2023 stepped down as chair of the Stream Corridor Restoration Committee of the Bayou Preservation Association.[29]

As if that were not enough to keep her busy, she is also the mother of three children and has a growing number of grandchildren, all of whom live within a mile of each other. She also mentors young people interested in working in conservation.

Mary Anne's legacy is still a work in progress as Houston rolls out resiliency and climate plans that call for an investment in green infrastructure. Her legacy certainly resides in the Texas Land Trust Council

Putting Down Roots volunteers planting prairie grasses, 2016. Courtesy Coastal Prairies Conservancy

that she helped grow and strengthen as it prepares a significant push for greater state investment in conservation lands and easements through the Texas State Conservation Funding Initiative. Mary Anne acknowledges it might take several sessions to secure the funding.

Her legacy also can be found in the thirty thousand–plus acres of the Coastal Prairie Conservancy's holdings and in the staff and volunteers she has nurtured over the twenty-four years she has served that organization. Mary Anne credits her talented staff and board of directors as key to the success of the conservancy and its projects. Her legacy will be felt with each rainstorm that becomes a flooding event and is slowed and eased by the prairies and creeks that she helped protect through her flood-zone-to-green-zone concepts. That legacy can also be witnessed each fall and winter as thousands of ducks, geese, and shorebirds return to the wetlands and tallgrass prairies she championed. Perhaps most importantly, it will live on in the awe and smiles of children encountering nature at the preserves, enjoying a conservancy education

Mary Anne (*right*) and Lydia Saldaña at the Katy Prairie filming for the first Audubon Texas Women in Conservation Terry Hershey Award, 2014. Courtesy Lydia Saldaña

or outreach program, or exploring one of the pocket prairies across the greater Houston area.

Throughout her long and meaningful career, Mary Anne has won numerous awards, including being part of the inaugural class of honorees for the Audubon Texas Women in Conservation Terry Hershey Award in 2014.

CONSERVING TEXAS' RANCHING AND NATURAL HERITAGE

Blair Calvert Fitzsimons

Blair Calvert Fitzsimons was born and raised in Texas and fondly remembers exploring the wilds of her grandparents' ranch as a child, casting seeds in the pasture with her grandmother. After high school, Blair attended Princeton University to study art history and English. She always loved the outdoors but did not discover the important relationship between land management and conservation until after she met and married Joseph Fitzsimons, who also came from a ranching family with deep roots in the Texas soil.[30] The two married just six months after meeting each other at a friend's wedding.

In 1989 the young couple moved to the Fitzsimons family's San Pedro Ranch in South Texas after Joseph assumed its management. Blair handled the day-to-day operations while Joseph practiced law in Carrizo

Blair and Joseph Fitzsimons, April 2022. Courtesy Texas Agricultural Land Trust

Springs. Before retiring and leaving the ranch in Joseph and Blair's capable hands, Joseph's father helped bring Allan Savory from Africa to train others on his holistic resource management. Allan had developed a grazing and land management system that mimicked natural herd movement and demonstrated how to rotate cattle from one pasture to the next so the grasses and soils could recover. Holistic management was not a technique that every rancher accepted at the time, but the Fitzsimonses saw it as a tool for protecting the arid grasslands found on their South Texas ranch. Learning and practicing the holistic management model was Blair's first introduction to land and water conservation. She eventually served on the board of Holistic Management International–Texas (formerly Holistic Resource Management of Texas).[31]

Blair and Joseph raised and homeschooled their three children, Fay, Jonny, and Kate, while living on the ranch. In our 2022 interview, Blair laughed as she stated, "Algebra I is what sent us packing to San Antonio in search of a more advanced education system."[32] While in San Antonio, Blair worked for the Bill and Melinda Gates Foundation to establish the $100 million Texas High School Project. She then went on to chair the Telecommunications Infrastructure Fund Board, a $1.5 billion fund established by the Texas legislature to award grants to rural schools, hospitals, and libraries for telecommunications infrastructure. While on the board, she became familiar with making policy and navigating the inner workings of the legislature.

Interested in finding solutions that would allow farmers and ranchers to maintain their lands as working properties to keep them in the family for future generations, Blair decided to join the American Farmland Trust. "To be honest," she remembered, "when I went to work for the Farmland Trust, I had no idea how a conservation easement worked or what a land trust did, but I knew how to build coalitions and get things done at the state capitol. More importantly, I embraced the idea that we needed to have tools in the toolbox to deal with land fragmentation. We needed a way to keep families on their lands. So, I learned."[33]

While in her role with the American Farmland Trust, Blair led the effort to pass the Texas Farm and Ranch Lands Conservation Program. After several rounds of negotiation, it was decided that the unfunded program would be managed by the Texas General Land Office. The program was awarded federal funding that allowed for seven grants to be issued in 2006 to conserve roughly ten thousand acres. The selected lands also protected millions of dollars' worth of surface and underground water.[34] In 2016 the Texas legislature transferred the oversight of the Texas Farm and Ranch Lands Conservation Program from the General Land Office to the Texas Parks and Wildlife Department.

The 2005 unfunded mandate that created the Texas Farm and Ranch Lands Conservation Program specifically focused on protecting agricultural lands. However, without funding, the agriculture community requested that a land trust be created to ensure that the intent of the bill would be honored. In 2005 Blair, with support from Bob McCan, rallied a stellar board of directors together from ranching and farming families and hunters to form the Texas Agricultural Land Trust, which was incorporated in 2006. Earlier that year, according to a land trust report, "Congress passe[d] legislation creating enhanced tax incentives for donated conservation easements. These new incentives, which include special provisions for full-time farmers and ranchers, generate significant demand for an ag land trust in Texas."[35]

With Blair as the executive director, the new land trust focused on working ranches and farms in which families wanted to conserve the wildlife, water, and lands for future generations. Water conservation is still a primary focus for the Texas Agricultural Land Trust; without water, neither productive crops nor healthy livestock or wildlife would exist. Right out of the gate, the Texas Agricultural Land Trust successfully secured several conservation easements in the Davis Mountains and the Panhandle with landowners who were willing to share their stories and encouragement with other landowners.

Blair shared that part of the Texas Agricultural Land Trust's success stems from landowners helping each other to overcome hesitations and find the facts they need. The Texas Agricultural Land Trust utilized and continues to support the Texas A&M Texas Land Trends study, which bolsters the case for conservation on farms, ranches, and timberland. In 2003, when Blair was at the American Farmland Trust, she helped raise money for the longitudinal study, which tracks changes in land ownership, fragmentation, and the impact of those changes on water, agriculture, and wildlife. "The data coming out of Texas A&M University showed we were losing our ag lands at a faster rate than any other state in the country. And it's still true today," Blair said. "It was the first time anybody had really taken a hard look at the land loss, specifically ag land loss, and what the implications were."[36]

In 2014 US Representatives Lloyd Doggett, Lamar Smith, and Michael McCaul joined their congressional colleagues to cosponsor HR 2807 to make permanent tax incentives for land conservation. With support from Blair and land trust leaders from around the country, the Conservation Easement Incentive Act passed. That act, partnered with a restructuring of the inheritance tax laws, allows property owners who are land rich and cash poor to better access conservation tools, thus enabling a greater diversity of large and small ranches and farms to participate.[37] Prior to this combination of changes, families of color, who typically own smaller farms and ranches, were left out of the benefits of conservation easements.

That same year, Blair and her team kicked off an awareness campaign to highlight the unique position that private landowners hold in the conservation matrix. No Land, No Water was designed to educate landowners, elected officials, and private citizens on why they should care about farm and ranch preservation and the alarming rate of lands lost each year. As Blair said, "We are winning a race we don't want to win."[38] Once that land is withdrawn from food and fiber production and built on, it loses the ability to absorb water into aquifers, streams, rivers, and lakes. In Texas, 142 million acres are privately owned; most of that acreage is found in farms, ranches, or forests.[39]

In a 2017 article for the *San Antonio Express-News*, Blair stated,

> By 2060, Texas's population will grow to 40–50 million people. If we are to ensure sufficient water supplies for these future populations, we need to act today to protect the lands that will help supply the water that goes into those pipes and pumps provided for by the state's water plan. It's no secret that private landowners have been supplying food, fiber, and economic prosperity to our state for centuries. But we must not forget that their success and tenure on the land is also responsible for much of the clean drinking water that

Mike Conaway receiving recognition for his support in passing the 2018 Farm Bill, 2019. *Left to right:* Josh Maxwell, Wendy Jackson, Representative Mike Conaway, Blair Fitzsimons, and Romey Swanson. Courtesy Texas Agricultural Land Trust

> sustains our growing population. Whether you live in downtown Dallas or the outskirts of Alpine, you likely feel a great deal of pride in the heritage and cultural values that make you a Texan. If you are a Texan for Water, you are a Texan for Land. *No Land, No Water*. In his keynote address to the Texas Wildlife Association's annual conference last July, [Texas House] Speaker Joe Straus commented that ". . . providing water is not all about pipes and pumps. It's also about protecting our aquifers and rivers, and we need private stewardship to do so. Almost every drop of water in this state falls first on private property."[40]

Blair's legacy continues to flourish in the Texas Agricultural Land Trust, from which she retired in 2020. From 2007 to 2020, the trust negotiated thirty-three conservation easements to protect more than 236,000 acres of land in twenty-four counties. Those lands provide approximately 178,000 acre-feet of water each year and protect 569 species of plants and animals, helping stem the tide of land fragmentation by holding some of the largest easements in the state.[41]

Blair's legacy can also be found in the research and legislation she championed, the people she mentored or employed, and, most importantly, her family. Firm believers in positive stewardship and avid hunters, Blair and Joseph have managed their lands as working ranches that also provide habitat for wildlife. One of the many ways Joseph benefited conservation in the state was through his appointment by Governor Rick Perry in 2001 to the Texas Parks and Wildlife Commission. He served a seven-year term; four of those years he served as chair. Joseph cofounded and chaired the Texas Coalition for Conservation to secure legislation and a subsequent constitutional amendment that designated the Sporting Goods Sales Tax to permanently fund state parks and historic sites. In 2023 he chaired the Texas Coalition for State Parks to secure $1 billion in the State Parks Centennial Fund that will go toward purchasing new state parks. Since gaining access to the fund, the state park system has added more than 7,203 acres by June 2025.

Blair's own story reflects what so many generational landowners have experienced. One of four cousins who now operate the Calvert Brothers Ranch, Blair reminisced, "Our annual family gatherings started in 1976 with a small campout with our dads. Now, there's about two dozen of us who gather each year for a treasured family tradition."[42] Thanks to the land trust she established, she and her family had options to protect their land for future generations and become part of the conservation matrix.

As has happened to many properties across Texas and the US, as the Calvert Brothers Ranch passed down from generation to generation, it was divided into ever-smaller pieces. With increasing operating costs and estate taxes, along with rising land values, many families face difficult decisions about the fate of family land. Blair opted to wait until her retirement in 2020 to seriously broach and pursue the idea of a conservation easement with the family. By 2021 the family began working with the Texas Agricultural Land Trust to negotiate and close a conservation easement on their land. The easement forever protects 4,718 acres on the banks of the Leona River, which is shaded by enormous oak trees and home to a variety of riparian species and native wildlife.[43]

Blair at the San Pedro Ranch, 2020. Photo by John Dyer. Courtesy Texas Agricultural Land Trust

In 2016 Blair and Joseph were honored with the Texas Parks and Wildlife Department's Lone Star Land Steward award, and shortly thereafter they won the Harvey Weil Conservationist of the Year award. In 2020 they were awarded the prestigious Leopold Conservation Award from the Sand County Foundation. In 2017 Blair was honored with the Audubon Texas Women in Conservation Terry Hershey Award.[44]

Journalist Pam LeBlanc documenting the Texas Water Safari paddling race on the Guadalupe River, 2018. Photo by Robert Youens. Courtesy of Pam LeBlanc

images. And Shannon Davies carefully curated authors focused on natural history, outdoor exploration, and species identification while in her role as editor at the University of Texas Press and as director of Texas A&M University Press.

We depend on storytellers to engage us to notice the nature that surrounds us every day or inspire us to seek the excitement of a distant, new place. They are an essential element in the conservation collective.

THE POETRY OF NATURE

Karle Wilson Baker (1878–1960)

Karle Wilson Baker was not only the most celebrated female poet, novelist, teacher, and essayist of the first half of the twentieth century, she was also an avid birder and naturalist. Her stories and poetry celebrated the natural world and inspired others to see it with their hearts as much as their minds.

At her family home in Nacogdoches, known as Tanglewood, Karle often sat at her window or in the yard to watch the birds. She poured her observations into a lovely book titled *The Birds of Tanglewood*, for which her daughter, Charlotte Baker Montgomery (1910–2009), provided the illustrations. Karle eloquently described the avian life of East Texas during the early 1900s in a way that was not scientific and heavy but instead was captivating, relatable, and poetic.[1]

As a renowned poet and writer, Karle was able to support herself before her marriage (not an easy thing for a female or male writer to do in those days). Born in Little Rock Arkansas, Karle attended university and worked as a teacher at Southwest Virginia Institute, a school for young women in Bristol, Virgina. She earned enough money to attend the University of Chicago. As she attended college, she began to establish herself as a writer. When Karle's father's business in Arkansas failed, he moved the family to Nacogdoches in hope of finding opportunities to rebuild. Money was tight, which

Karle Wilson Baker, 1919. Courtesy East Texas Research Center, Stephen F. Austin University

made Karle want to be self-reliant even more. While working as a teacher offered financial stability, it was also draining emotionally. Karle turned more and more to nature to recharge herself. When Karle's mother's health began to fail, Karle permanently joined the family in Nacogdoches.[2]

She quickly fell in love with East Texas. She also fell in love with a Nacogdoches banker and businessman, Thomas Ellis Baker. The couple had two children, Thomas Wilson and Charlotte. In 1910, the same year Charlotte was born, Karle's mother passed away. True to the traditions and social expectations of the time, Karle was a devoted mother, wife, and daughter who put caring for the family first and everything else second. Thankfully, her husband cherished her brilliance and encouraged her to remain loyal to her poetic nature and continue writing. It is reported that he also enjoyed the outdoors and bird-watching with the family.

In 1916 *The Yale Review* published seven of Karle's poems, all of which had nature themes. That same year, her father, W. T. Wilson, was struck by a train in Lufkin and killed. After her father's passing the Wilson's moved into Tanglewood. Martha Anne Turner notes in her book, *Old Nacogdoches in the Jazz Age*, "Bird watching became an absorbing avocation during these nostalgic years. Evidently the entire family enjoyed the hobby, as they deliberately left the thickly-shaded grounds and underbrush and tangles of grapevines of Tanglewood undisturbed, to the scandal of the town."[3]

As early as 1906, Karle is on record as advocating strongly for a state school to be established in Nacogdoches. The goal was finally achieved in 1924. The new president of Stephen F. Austin State Teachers College ignored the state rule that no married woman should be allowed to teach at a state university and offered Karle a position in the English Department. After all, that is what you do when you have one of the greatest living poets residing across the street from the newly formed university. During her time at Stephen F. Austin (between 1924 and 1934), Karle was prolific, producing poetry, short stories, and two children's readers.[4]

Karle was one of the few women of her time to be published frequently in *Harper's* and *The Yale Review*—in addition to a myriad of other national publications. She was nominated for a Pulitzer Prize in 1931 for her compilation of poetry *Dreamers on Horseback*.[5]

Karle's poetry and stories tantalized the senses and begged people to drink in the elements of nature surrounding them. In addition to writing beautiful poems and stories that celebrated nature, the ways of the heart, and daily life in Texas, Karle was also a charter member of the Poetry Society of Texas, Philosophical Society of Texas, and Texas Institute of Letters. She was the third person honored as a fellow to the Texas Institute of Letters, just after J. Frank Dobie and Walter Prescott Webb. Like other authors of her time, Karle witnessed unpredicted changes to the natural world including the felling of the forests, fencing of the prairies, loss of species and the expanding petroleum kingdom. She often wrote of the loss of nature and the loss of Native Americans and their traditions of understanding the land and its creatures.

While Karle's work was published frequently in national publications, it also appeared in local Texas newspapers and magazines. She earned additional income from speaking engagements across the country, with frequent stops around Texas. Because of the spelling of her name, organizers were often confused when a woman appeared to speak instead

Karle Wilson Baker with a member of the Alabama-Coushatta Nation, 1930s. Courtesy East Texas Research Center, Stephen F. Austin University

of a man. Her speeches often pulled at the heartstrings to inspire a deeper connection with the natural world. In 1923 she was one of the headliners, along with Governor Pat Neff, at the annual conference of the Texas Federation of Women's Clubs, where the governor advocated for a state park system.

Karle's legacy is immortalized in her poetry and stories that connect the reader with the ethereal exquisiteness of nature. Her essays, books, and poetry opened the door for other women—and men—to tell the story of nature in a way that evokes compassion and a desire to care for it. Her legacy also lives on in the Texas Institute of Letters, which remains dedicated to promoting authors who tell the story of Texas' land, water, wildlife, and people.

Her daughter, Charlotte, followed in her mother's footsteps and became an author of over twenty children's books. Charlotte, also an avid birder and naturalist, was a member of the Texas Ornithological Society, and founded the Nacogdoches Human Society. In 1983 she honored her mother by donating sixteen acres of land to the City of Nacogdoches, with a later additional gift to the Texas Land Conservancy. Karle dearly loved the woods behind her second home off Rusk Street and walked to the woods frequently to immerse herself in nature and find inspiration. Banita Creek Nature Preserve is worth a visit to listen for the thrushes at dusk.

At the corner of Hughes and Mound Streets, near Karle's beloved home of Tanglewood in Nacogdoches, there is a statue dedicated to her. It features one of her poems:

Through Tanglewood the thrushes trip,
As brown as any clod,
But in their spotted throats are hung
The vesper-bells of God.
And I know little secret truths,
And hidden things of good,
Since I have heard the thrushes sing
At dusk, in Tanglewood.

A NATURAL STORYTELLER

Lydia Saldaña

Lydia Saldaña told and improved the story of the Texas Parks and Wildlife Department for more than twenty years. During that time, she immersed herself in all aspects and functions of the agency so she could stitch together the story of conservation in a new, relatable, and relevant way. However, she did not start out as a "nature lover." When asked whether being outdoors was a large part of her childhood, she replied strongly, "Not. At. All."[6]

Lydia Saldaña at McKinney Falls State Park, 2010. Courtesy Texas Parks and Wildlife Department

Lydia's father was a construction contractor in San Antonio who was ready to be home indoors on his days off. Growing up, Lydia was more interested in watching *60 Minutes* or reading the newspaper than hanging out in the outdoors with her friends. In high school, she was mentored by the first Hispanic female broadcaster in the San Antonio market. Her mentor encouraged her to follow her dreams to become a television news journalist. She also encouraged her to attend college; Lydia was the first person in her family to attend.

However, in 1981, she dropped out of Trinity University in San Antonio after landing a job with KLBJ News Radio in Austin. In 1982 she joined the team at KXAN-TV to become one of the first women of color in the Austin market. While she was at KXAN-TV, a colleague urged her to finish her degree so she could be competitive in the television journalism field. In typical Lydia fashion, she completed her degree in journalism at the University of Texas at Austin while working a full-time job.

In 1988 she landed a job at WFAA-TV in Dallas. She described her time there as "glamorous." As a general assignment reporter, she attended astonishing events, met interesting people, and personally crafted each story she was given. Her husband was still working in Austin, and Lydia admitted the long-distance relationship was hard. In 1990 she put family first and returned to Austin. But she was out of a job.

Feeling like she might have made a mistake walking away from a prestigious television station in Dallas, she applied for a video production manager position at the Texas Parks and Wildlife Department. While waiting in the lobby at the department's headquarters to interview, she

Lydia Saldaña with photographer Jim McKenna, both with KTVV (now KXAN), at the state capitol for the replacing of the Goddess of Liberty, 1986. Courtesy Lydia Saldaña

picked up a brochure titled *Now You've Killed Your First Deer.* She thought, "Where the hell am I, and what am I doing here?" The poorly written brochure on how to field dress a deer was about to get a serious upgrade—along with all of the agency's other marketing pieces.

Andrew Sansom joined the department later that same year as executive director. Andrew had been the executive director for The Nature Conservancy and had served as a special assistant to US Interior Secretary Rogers Morton. Lydia credits Andrew for making the agency one of the most professional in the country and recruiting and lifting up women in the department. "He saw something in me that I didn't even know was there," she shared in a 2020 interview.[7]

She also shared that shortly after starting at the department, she became pregnant with her second daughter. Raising two daughters and commuting to Austin from Dripping Springs was not easy.

Just a year into joining the department, Lydia was promoted to Media Services Division chief, which allowed her to weave her magic into the agency's internal as well as external communications. Because she was such a novice at every outdoor experience herself, she wanted to tell that story and take people on a journey of discovery through activities they could see themselves doing at some point in the future. "In the mid-'90s," she recalled, "I really started hammering the idea that we had to stop preaching to the choir. It's a good choir, but we needed to reach more Texans. Not just the ones who already hunted and fished. One of the first statewide campaigns we tried was called Don't Be a Pain in the Boat. It was a multiagency collaboration designed specifically to reach urban and suburban audiences who were boating but maybe not fishing."[8]

The campaign made a splash and helped propel Lydia to become the first woman Communications Division director in 1996. As early as 1994, the Texas Parks and Wildlife Department was crafting a plan in response to the changing demographics of Texas and waning interest in hunting and fishing. The data showed that people were becoming more interested in pursuits such as bird-watching, camping, and hiking. Lydia and her team took the opportunity to explore stories and events that would engage urban audiences with Texas' best-kept secret: the state parks. In those days, getting a campsite at a state park was no problem; visitors just called the park or showed up. Today, that's not the case.[9]

Lydia and daughter Shannon on Take Your Daughter to Work Day, on location at Palo Duro Canyon State Park, 1993. Courtesy Lydia Saldaña

Lydia always kept a focus on engaging people of color with the programs offered by the department, which had not always pursued that goal. She often said that "words and images matter." Lydia insisted that the agency's photographers and videographers include people of color when they were in the field or staging a photo shoot. Admittedly, since most of the people working at the department at the time were white males, it was hard to find a Hispanic, Asian, or African American family at a moment's notice. (Texas Parks and Wildlife families and friends were often used in photos as unpaid models to keep expenses low.) Eventually, she had to find a handful of models to fill the void.

As the communications director, she oversaw a team of seventy people whom she repeatedly credits for her own success at the agency. In 1998 she hired Darcy Bontempo as marketing director. Darcy, who had worked for GSD&M advertising in Austin, stated that one of the best things about working with Lydia was that "she really cared about every person on her team and totally had their backs on the good days and bad. If someone on the team had a solid idea that could improve a current system or even be a new campaign, Lydia was not afraid to try it out.

Lydia with daughter Rachel at Enchanted Rock, 1994. Courtesy Lydia Saldaña

She was always looking for fresh ideas and ways to introduce new people to all aspects of the Texas Parks and Wildlife Department. Surrounding herself with a great team was a large part of that. She believed in our creativity and built a supportive environment for us to reach beyond people already using the parks or hunting and fishing." With the population of Texas exploding in the 2000s, reaching those audiences became more important than ever.[10]

Lydia was on the front end of deploying a multidiscipline approach to reaching audiences, including transitioning to a website, videos, and news stories, as well as preparing for the looming social media trend. Her team also focused on nature tourism, outreach, and education as well as *Texas Parks and Wildlife Magazine* and the radio show. Each department represented a touchpoint for various demographics to engage with the agency and nature. Lydia also raised the bar of professional development through media training, not just for her team but throughout the agency and other nonprofits. She was able to share her expertise in public relations with students at Texas State University while she earned her master's degree in mass communications in 2011.

One of the largest-scale campaigns ever launched by the agency was developed in response to a survey that showed that the average park user or individual buying a hunting or fishing license was around forty-four years old, white, and male. In 2006 data were also showing that children were spending between seven and eleven hours per day indoors interacting with some form of media and only minutes per day playing in nature. Something had to be done.

Under the direction of Lydia, Darcy Bontempo, and the agency's conservation education manager, Nancy Herron, they pulled together a team to generate ideas. The Life's Better Outside campaign was introduced. Most conservation-minded people think the words *outdoors* and *outside* are interchangeable. But that is not the case for all demographics. And for Lydia and her team, the power of each word mattered. They had to fight for the word *outside*, as it implies nature that is nearby, such as a front yard, local park, or nature center, something that is more tangible for an urban dweller. It was a tremendous shift in mindset for the Texas Parks and Wildlife Department but had ripple effects in the rest of the conservation community, encouraging them to value all the spaces where people engaged with nature. The challenges were overcome, and the campaign's name became the official tagline of the agency. The concept was supported by a variety of emerging programs focused on urban engagement.[11]

The project that generated the most satisfaction for Lydia was a ten-year campaign called Texas: The State of Water. It kicked off with the July 2002 issue of *Texas Parks and Wildlife Magazine*, which was filled with articles about Texas' creeks, rivers, lakes, and coast. It was followed by an award-winning program produced by Lydia's team that aired on PBS. The

Texas Parks and Wildlife Magazine, 2002. Courtesy Texas Parks and Wildlife Department

messaging was informative and always had a positive call to action for citizens to make changes in their daily lives to conserve and protect water. Every July issue of the magazine from 2002 to 2011 focused on water, which was timely as Texas saw one of the worst droughts and some of the hottest summers on record from 2009 to 2012.[12]

Lydia stated, "So often, press releases and media stories are in reaction to something or of an immediate nature. I was really proud to support this ten-year project that focused on such an important issue. And I love it when people tell me that they used the videos in their classrooms or saved the magazines to refer to again and again."[13] (When I was the program director at Camp Fire Central Texas, the kids watched these videos at the thirty-six different schools where we had after-school or summer programs. We even developed hands-on lessons that went with each video.)

In 2013, recently remarried, Lydia decided it was time to explore new trails and retired from the department. Her retirement lasted about two days before Anne Brown, executive director of the Texas Parks and Wildlife Foundation, reached out to recruit her. Lydia kept her promise to her husband that they would travel and enjoy adventures together; however, in between playing on the beaches of Greece and Galveston, she accumulated an impressive roster of conservation organizations to which she provides communications consulting: the Borderlands Research Institute, Western Association of Fish and Wildlife Agencies, Harte

Lydia conducting interviews with Carter Smith (*left*) and Andrew Sansom (*right*) for the Audubon Texas women and conservation video, 2018. Courtesy Texas Parks and Wildlife Department

Research Institute, Rolling Plains Quail Research Foundation, and Friends of the Fort Worth Nature Center and Refuge, to name a few.

Lydia noted that one of the highlights of her consulting work was producing videos for the Audubon Texas Women in Conservation Terry Hershey Award. She revealed, "These women are so inspiring, and it is a thrill to find the thread that laces all their efforts together into a three-minute video segment."[14]

Telling the story of nature at the premier outdoor agency in the second-largest state in the country during a time of massive growth leaves a pretty large legacy. Lydia pushed for the agency to be inclusive beyond the "hook-and-bullet" crowd and actively seek out welcoming engagement opportunities for all demographics. Her attention to messaging and shrewd use of images to not only tell the story of nature but encourage people to care about it have inspired the marketing and communications teams at other parks and wildlife agencies and nonprofits across the country. Her reach extends to millions of Texans and visitors who have discovered that life is better outside.

Texas Parks and Wildlife executive team, 2012. Courtesy Texas Parks and Wildlife Department

CREATING STATEWIDE VOLUNTEER NETWORKS

Thea Platz

Thea Platz grew up in a military family that moved frequently. In our 2022 interview, she shared that her grandparents' homes in northwest Arkansas, where the family spent summers and holidays, were places of stability. Her father's parents had a farm in the country and a Sunday house in town. Her grandfather, a self-taught naturalist, introduced Thea to the world of plants. Her grandmother was active in their church and stockpiled boxes of clothing and household items that she would deploy if someone was in need. Thea recalled that if someone's house burned or some other tragedy occurred, people in the community knew they would be cared for at her grandparents' home. "My grandmother taught me what it meant to give service and care."[1]

Her other grandparents had a vegetable garden and flowers blooming throughout their yard. The floral theme continued inside the home, where her grandmother's collection of doilies, vases, and teacups brightened the house. Thea reflected that this grandmother imparted the importance of manners and instructed her on how to politely and gracefully move in a variety of social circles.

While her family was stationed in faraway places like Hawaii, Thea said, she always found a tree or bush where she could hang out, either by herself or with friends. The youngest of four siblings, the rest of whom were much older, Thea often sought out nature for entertainment and continuity as the family relocated from base to base. The trees became her sanctuary.

Thea married in 1975 and welcomed her only child, William, into the world a decade later in 1985. The young family lived in San Antonio, where she attended college and studied business while also volunteering at her church. At church, she saw how disconnected children had become from where their food came from. She worked with others to install garden plots and created a curriculum in Bible school that connected Bible lessons to growing, harvesting, and cooking food. She remembers watching children's self-confidence improve as they learned the basics of self-reliance. Recognizing her organizational talents and ability to connect with the congregation, the church elected Thea to become chair of the board of directors. She was the second woman and youngest person to serve in that role.

Always looking for nature escapes, Thea signed up to take a hike with naturalist Walt Schuman, who was clad in traditional Bavarian costume, complete with lederhosen. The two connected over plants, and Thea soon volunteered to become an interpretive guide at the Friedrich Wilderness Park on the west side of San Antonio. Having recently been through a divorce, Thea recalled that becoming a guide gave her a sense of happiness and fulfillment during those trying times of adjusting to being a single working mother.

While volunteering, Thea met Debbie Reid, who was hired by the City of San Antonio Parks and Recreation Department to serve as the nature preserve coordinator at the Friedrich Wilderness Park. Debbie needed volunteers who had a working knowledge of the plants, animals, and topography of the park. She had worked extensively with the Texas Master Gardener Program and expressed to Thea and others that she wished she had a similar volunteer army that was focused on nature. The seeds were planted for what was first called Natural Initiatives: Conserving Urban Wildlife and Resources.

The concept of the initiative was to expand training for the Friedrich interpretive guides and the park's volunteers to include general natural resource information for Bexar County and to create a volunteer corps that could cover multiple sites and agencies rather than having to recruit and train volunteers for each location. In this way, the corps could be deployed for activities such as removing invasive plants, maintaining trails, assisting with plant and wildlife surveys, leading educational programs, or just helping where needed.

Being understaffed and underfunded is not unique to the Friedrich Wilderness Park. When the small group instructing the volunteers saw a greater need for well-trained natural resource volunteers beyond the county, they formed the Texas Master Naturalist Program. Thea employed her organizational and administrative skills to craft the group's mission, draft its bylaws, create its training manual, and devise its administrative systems to allow the project to scale.

Meanwhile, Debbie connected with professionals from various organizations and agencies such as the Texas Parks and Wildlife Department, Texas Forest Service, and Texas Agricultural Extension Service to look for funding and support. Once the nonprofit was formed, Thea was elected as the first president and served in that capacity for six years. Under her leadership, the organization grew in capacity from a single chapter in San Antonio to a statewide volunteer army.

From 1997, when it was officially formed, to 1999, all duties were completed by volunteers. The small community grew from around forty members to two hundred "who preach the wonders of nature in South Texas and implement programs enhancing environments in parks, schools, neighborhoods, and other areas around the region."[2]

Thea described that time: "We were developing the lessons and doing the trainings in real time. The handouts for the training binder would still be warm from the copy machine as we passed them out to the trainees. We had really talented people involved in writing the lessons, including Judit Green and Rufus Stephens, both urban biologists from Texas Parks and Wildlife; Vickie Marshall, who was a naturalist with the City of San Antonio; and others. Vicki Worthington would edit the copy and design it on her computer. Having access to a computer in those days was a big deal. Not everyone had them."[3]

In 1999, after partnering with the Texas Parks and Wildlife Department and Texas A&M AgriLife Extension Service, the organization was able to hire Michelle Haggerty as its first statewide program coordinator. Michelle was stationed at the Texas Parks and Wildlife Department in the nongame and rare species program under the direction of John Herron. Having a statewide coordinator allowed the organization to once again scale its capacity to reach a growing need around the state to connect volunteers with meaningful environmentally based projects. With the support of Texas A&M AgriLife, Mary Pearl Meuth joined the team in 2014.

In 2000 the organization won the President's Award from the Wildlife Management Institute. In 2001 the Texas Master Naturalist Program, San

Thea Platz receiving an award from the Texas Parks and Wildlife Department and the Texas Master Naturalists for her years of service, 2004. Courtesy Michelle Haggerty

Antonio River Authority, and Government Canyon State Natural Area were among the twelve winners of the Texas Natural Resources Conservation Commission's Environmental Excellence Award from the Texas governor's office.[4] Thea was the first Texas Master Naturalist to be recognized for giving five thousand volunteer hours and received the President of the United States' most prestigious volunteer honor, the Call to Service Award.

In 1997 Thea joined a citizens' advisory board for the City of San Antonio Parks and Recreation Department. In that capacity, she connected again with Chairman Walt Schumann, archeologist Herb Uecker, and others who were working to designate the Walker Ranch as a park. The property was located near the confluence of Panther Springs and Salado Creek. Archaeological research conducted by the University of Texas–San Antonio revealed the site had long been used by prehistoric people as a hunting camp and gathering point. According to the City of San Antonio's website, "Walker Ranch was added to the National Register of Historic Places in 1975 and, in 1995, was listed as one of the most endangered historic places in the state."[5]

Thea and the team not only helped raise money and awareness; they also united the City of San Antonio, Bexar County, and the San Antonio River Authority to establish the park—later named Walker Ranch Historic Landmark Park. It was the first project where all three entities pooled their resources to create a park. Thea recalled, "I needed some paperwork signed to apply for a grant. The deadline was fast approaching, so I drove to former Mayor Lila Cockrell's office at the Parks Board. [Lila served as the first female mayor of San Antonio from 1975 to 1981 and was elected again in 1989, serving until 1991; she also served on the San Antonio Parks Foundation from 1989 to 2013.] Lila was packing her office to move to another workplace. I jumped in to help pack her artwork and objects, got her to sign the papers, and then raced out the door to get them in by the deadline."[6]

That grant was from the Kronkosky Charitable Foundation to build a fully accessible playground at the park for children with limited mobility. Thea and the team engaged Churchill High School students to install the playground structures. The park was designed as an outdoor learning center, complete with an open pavilion for a classroom. Thea remembered, "While involved with Walker Park, I really started to feel the ripple effect. We were able to get Joy Fisher Hein, a local—but not yet well-known—artist to design the interpretive panels. During installation of the panels, a café owner drove by and saw the students working on the playground; he brought them lunch every day from his café. Then the president of the school board of the Northeast Independent School District got involved. He let me know he wanted a nature-based education program at the school district and asked if I would be interested. They hired me, and we used the park as the outdoor classroom."[7]

Thea became the first director for outdoor education at the North East Independent School District; however, she credits coworker Tom Anderson with launching an overnight camping program at the district. They contracted with John Knox Ranch, a Presbyterian-run summer

Children canoeing at the John Knox Ranch camp on the Blanco River, 2019. Courtesy Henry Owen, Executive Director, John Knox Ranch

camp on the Blanco River, to host overnight camps during the school year. Thea developed the site-specific curriculum that checked all the boxes for the Texas Essential Knowledge and Skills requirements, connected children with nature, and helped them develop interpersonal and leadership skills.

The program would host two camp-outs a week, with one set of fifth graders arriving on Monday and leaving Wednesday morning and the second set arriving Wednesday afternoon. The rustic camp on the banks of the Blanco River near Fisher, Texas, was the perfect place to immerse children in nature, often for the first time. All the children in the program came from socioeconomically disadvantaged communities and Title One schools.

Thea shared a story about a blind student who attended the program with his father. The child had spent his entire life walking the flat surfaces of sidewalks, homes, schools, malls, and the like. When he encountered the trail's rough surface and its incline, he touched each piece with his walking cane, erupting in "peals of laughter." His face lit up with awe as he encountered each new natural wonder. Thea stayed by his side for most of the program, including when the children went fishing with old cane poles and worms. The blind child cast his lure into the water and, moments later, hauled out a large fish. Then nine more after that. His sensitive hands could feel the tiniest nibble on the lure, allowing him to react instantly. Thea could see his father beaming with joy as he saw his son learning a lifelong activity that they could enjoy together. And for the first time, the child was the best at something physical. He caught the most—and largest—fish and was acknowledged in the closing ceremony. Thea added that it was moments like this that kept her and the program going for twenty-five years.

In 2002 Thea added a day fieldtrip program for the district's third-grade classes. The third-grade program included parents and was hosted at Honey Creek State Natural Area. One of the projects Thea developed had the children create seed balls. Under the supervision of a state park's interpretive ranger, Craig Hensley, the Monday day campers made the seed balls, and the Wednesday campers cast them over the prairie restoration area. Over the course of the fifteen years of the program, the children, parents, volunteers, and park staff could see the impact of each of the projects. The encounters with nature gave families a connection with the parks that few families were able to experience. By Thea's design, it gave them a sense of place.

In 2015 a flood severely damaged the John Knox Ranch, and the overnight program came to an end. In 2016 and 2017 much of the Title One funding for out-of-school programming was cut at the federal level. Thea continued the day programs for third and fifth graders, which engaged fifteen hundred children each year. Over two decades, Thea's outdoor program hosted more than thirty-seven thousand children, approximately as many parents, and countless volunteers and programming staff.

In addition to planning, organizing, and executing the camp-outs and trips, Thea was a member of the statewide team that created the first Environmental Education Plan in Texas. She engaged with other teachers to encourage them to take Project Wild continuing education training so they could include more environmental education in the classroom, and

Texas Children in Nature San Antonio regional collaborative, 2015. Koy Coffer (kneeling, center) being presented with a rock in recognition of "rock solid leadership" as she passed the duties to Thea Platz. Courtesy Jennifer Bristol

she constantly supplied interested educators with books, ideas, supplies, and information to enhance lessons.

In 2010 Thea attended the first Children in Nature conference, hosted and organized by the Texas Parks and Wildlife Department's Nancy Herron, Kiki Corry, and Carter Smith. At the conference, Koy Coffer, education programs specialist with the Texas Wildlife Association, agreed to become the regional leader for the movement in San Antonio, with support from Thea and Judit Green. The three women created a network within Bexar County that consisted of conservation organizations, environmental educators, nature centers, parks departments, health care providers, and concerned individuals. Eventually, Thea assumed the role of regional leader as Koy's and Judit's jobs and families demanded more of their time.

The influences of Texas Children in Nature of San Antonio can be felt across the region. In 2015 San Antonio's Museum for Kids opened the DoSeum, which included a natural play space for children. In 2013 the museum brought in Richard Louv, author of *Last Child in the Woods: Saving*

Our Children from Nature-Deficit Disorder and cofounder of the national Children and Nature Network, to kick off its capital campaign.

In 2016 the San Antonio Botanical Garden purchased four acres of adjacent homes that were in disrepair and repurposed the land for the Family Adventure Garden, which opened in 2018. Thea was part of the botanical garden's advisory team that visited similar out-of-state garden projects, researched programs, and selected items to be included in the garden. The garden was intentionally designed to provide children with a place to roll down a hill, splash in a creek, build a fort out of sticks, climb boulders and logs, and create games on their own in a safe environment.

In 2017 the San Antonio Zoo, a charter member of the Texas Children in Nature Network, built and opened the first fully designed nature-based zoo preschool in the state. The Will Smith Zoo School allows children to spend 70 percent of their day outdoors, learning through playing in nature. The zoo school was a result of what Thea called the "ripple effect," as connections made at the 2015 National Children and Nature Conference hosted in Bastrop, Texas, culminated into an award-winning collaboration.

Because of the strength of the local collaborative and the get-it-done attitude of the network's members, especially Thea, the group was selected to host the first statewide Texas Children in Nature Network conference in 2017. Thea and Susan Campbell spearheaded the planning team. Susan was the education coordinator for Phil Hardberger Park and a Project Wild trainer who was never shy about asking others to lend a hand when needed. The conference was a huge success and a great place to showcase the variety of innovative projects and programs to connect more children and families with nature in San Antonio.

Following the conference, Thea and Susan led the charge to win the competitive Cities Connecting Children to Nature grant from the National League of Cities. San Antonio became the second city in the state to be awarded the innovative grant and compete for additional funding. Ultimately, the city's parks department took over as the lead, focusing on connecting underserved communities with local parks and nature.

When asked what she thinks is the most pressing issues facing Texas in the realm of conservation, Thea replied, "Land fragmentation is a

Thea Platz (*left*) and Bonnie Conner, president emeritus, Phil Hardberger Park Conservancy, at the Phil Hardberger Park Conservancy Nature Heroes Award celebration, 2018. Courtesy Jennifer Bristol

big one. But we also have an equity issue. Agencies keep talking about it, but they need to start diversifying their hiring. There is a delta between who visits the parks and the staff who work at the parks. We need to address that quickly. If we are focused only on recruiting, retaining, and reactivating people already interested in hunting and fishing or the outdoors in general, then we are missing the people who need nature the most. I hear a lot that kids need nature, but nature needs them, too. Nature needs kids to grow into adults who care about our parks, care about where our water comes from, care about the soil that grows our food. Nature needs them to become parents who demand that schools keep offering outdoor programs, agriculture curriculums, and gardens at schools."[8]

Thea added that she feels there is a need for more awe in children's and adult's lives. For Thea, awe comes in many forms, but nothing is as humbling and centering—or even as inspiring—as the connection with nature. "So much competes for our attention today. Nature is the great

from the Texas Parks and Wildlife Department to become its own 501(c)3, she was named one of its first board members.

Sadly, because of budget cuts in the district and lack of funding from the state, the twenty-five-year outdoor program and Thea's position at the North East Independent School District no longer existed as of spring 2023. Thea shared that while this is upsetting, she also looks forward to spending more time with her grandson and has faith that good things will fill her future. Thea's legacy lives on in each of the thousands of volunteers, children, parents, and teachers she inspired to discover the awe and wonder of nature.

THE MOVEMENT TO RECONNECT CHILDREN WITH NATURE

In the early 2000s research from the parks and recreation departments of various states showed that average park users, hunters, anglers, and other outdoor enthusiasts were approaching their mid-forties. More concerning, the research revealed that fewer and fewer children were participating in those pursuits, sounding an alarm about the future of parks and other outdoor interests. However, those data remained siloed in the various state parks and recreation departments for many years. Around this time, the Texas Parks and Wildlife Department was beginning to formulate its own ideas about how to engage more Texas youth, but something seemed to be lacking.

Nancy Herron, Lydia Saldaña, Darlene Lewis, Darcy Bontempo, and others at the Texas Parks and Wildlife Department were trying various methods to reach new audiences and spark an interest in the outdoors. Nancy was engaged in outreach via newsletters and articles and was tapped to spearhead the team to develop the agency's first website. Technology

Amanda Ross showing a Camp Fire program participant the fruit of a prickly pear cactus, 2013. Photo by Jennifer Bristol

was evolving quickly in the early 2000s, and she saw the website as a means to help the agency remain relevant.

With the agency's support, she also created its first live webcasts. "I thought we might have five hundred people sign up," she remembers. "We had five thousand for the first one. And it kept growing. Then I connected with NASA so we could increase our bandwidth. That was very exciting—and scary. But it took our message that 'life's better outside' directly into classrooms for the first time. And we always had kids be part of each webcast."[12]

The innovative idea received national attention, and Nancy was asked to be part of a NASA think tank. In our 2019 interview she shared her vulnerability about refraining from accepting the invitation: "One of my biggest regrets was pulling back from being a part of the think tank. I told myself I wasn't good enough, or smart enough, to be part of that. I talked myself out of going. Being a woman at an agency of primarily men, I guess I had gotten used to being in the background and not center stage. I saw other women pull back in the same way on other projects. We just weren't there yet. But it is a cautionary lesson that I feel I should share so other women don't do the same. We are good enough and should be at the table in all aspects."[13]

In hindsight, however, she reflected that if she had gone down that path, she might not have been so open to finding her way to the "children in nature" movement. Nancy continued working her way up the ladder at the Texas Parks and Wildlife Department and into the conservation education program within the agency.

Meanwhile, Molly Stevens was serving as the state development director for The Nature Conservancy and, later, the Environmental Defense Fund. Both jobs required a lot of travel. As her two children reached middle school age, she and her husband decided it was time to find something more local so she could be with the children more. In 2005 she accepted a position as the first woman executive director of Westcave Preserve (now Westcave Outdoor Discovery Center). At the time, Westcave was a small, thirty-acre preserve in western Travis County, with around five thousand visitors annually. The preserve's dynamic board, including Hayden Brooks, hoped to increase its footprint, grow its educational programs, and become more accessible to Austinites, especially children. They tasked Molly with achieving those goals—all of which took loads of funding and required an expanded network.

Enter author Richard Louv's landmark book *Last Child in the Woods: Saving Our Children from Nature-Deficit Disorder*. The 2005 book synthesized the data from those state park and recreation agencies, health departments, and education agencies, and it legitimized the sentiment so many people had that children were becoming further and further removed from nature. The book also challenged the reader to act with a sense of urgency and purpose to reconnect children and families with the outdoors. It galvanized the conservation and recreation communities and became a wake-up call to national, state, and local parks departments and schools to review their programs to consider how they could do more to engage youth. It also highlighted the growing childhood obesity and mental health epidemic, especially in middle school–age children.

The honesty in Louv's book required everyone in the conservation community to take a hard look at why and how the nation had gotten to this point so quickly. Why were adults allowing children to spend seven to eleven hours per day sitting indoors engaged with media and only minutes per day playing in nature? Why was an entire generation of parents afraid to take their kids camping? Why weren't schools taking kids on field trips to parks or teaching kids about nature? Why were we building cities for cars and not people? The finger of failure could not be pointed at one sector of society alone; it was a total systemwide breakdown.[14]

The dialogue and momentum to act began to grow. Hayden Brooks, an Austin-based architect and businessman, recalls reading *Last Child in the Woods* and having an overwhelming sense of resolve to raise awareness and create systemic changes. Hayden joined the board of directors of Westcave and saw an opportunity to utilize the preserve as a means to that end.[15]

In 2008 Molly Stevens, Hayden Brooks, Nancy Herron, and University of Texas Plan II Senior Research Fellow and Westcave Board Chair Lee Walker asked Richard Louv to come to Austin to discuss what they could do at the local and state levels. According to Molly Stevens, "Louv validated many of the ideas we already had and inspired us to move quickly. He shared what a handful of other communities were considering, but to be honest, there wasn't much of a road map to follow."[16]

From left: Hayden Brooks, Jennifer Bristol, Richard Louv, and Nancy Herron at the Children and Nature Network gathering in California, 2013. Courtesy Jennifer Bristol

Nancy Herron shared that after the meeting, she had a clear vision of what she wanted to do. "I knew Molly and Hayden were ready to focus on the local level in Austin. I wanted to utilize my position with Texas Parks and Wildlife to spearhead a statewide initiative. I also knew that if we were going to get the schools and health community involved, we'd need some solid research to back up our position."[17] With the support of her supervisors and Carter Smith, the new executive director of the Texas Parks and Wildlife Department, Nancy convinced the agency to invest in creating a research anthology and commissioned a new study.

In partnership with the Texas Parks and Wildlife Department, Hayden Brooks, Allen Cooper with the National Wildlife Federation, and a coalition of concerned nonprofits, a statewide effort emerged. They pushed for state legislation that would mirror the national effort to pass the No Child Left Inside Act. Texas State Representatives Donna Howard and Eliot Shapleigh introduced the Texas Partnership for Children in Nature Act, which directed the Texas Parks and Wildlife Department, Texas

Department of Education, Texas Department of State Health Services, and Texas Department of Agriculture, along with two members from nonprofits, to form a partnership. Their directive was to "promote the well-being of this state's children by providing children with opportunities to spend more time outdoors and to learn about the environment through experiential activities outdoors and formal and informal education concerning the environment."[18] The bill did not pass; however, a group of nine lawmakers who felt passionate about the effort wrote a letter to the governor requesting that the partnership be formed with the Texas Parks and Wildlife Department acting as the hub.

Carter Smith tapped Nancy Herron and the agency's outreach and education team to host a symposium to explore opportunities and create an action plan. The 2010 symposium took place at the LBJ Library in Austin and was attended by more than three hundred people. Participants, including myself, heard from speakers such as Joe Frost, the leading professor on play and childhood development at the University of Texas; Texas State Representative Donna Howard; Laura Huffman with The Nature Conservancy; Nancy Herron; and Jaime Gonzales from the Katy Prairie Conservancy.

On the second day of the symposium, attendees were asked to gather into groups by region to discuss how they would like to structure community action groups and determine their particular areas of focus for the following year. The groups fanned out under trees and on the lawn across the campus of the LBJ Library. Those pods of people formed the core of the subsequently created Texas Children in Nature Network. A leadership team that consisted of an individual from each region, plus Hayden Brooks, Allen Cooper, Nancy Herron, Margaret Russell, and Margaret Lamar, was formed to weave the statewide network together.

With leadership in place and actionable items on the horizon, Nancy Herron championed a group of eighty individuals to draft the statewide strategic plan.

Using the research anthology, the Texas Parks and Wildlife Department's marketing team and Texas Children in Nature leadership created the top ten reasons kids need nature. The list was incorporated into marketing pieces that utilized the unified message, "Studies show that kids who play and learn outside are healthier, happier and smarter."[19]

Healthier, Happier, Smarter poster for Texas Children in Nature. Courtesy Texas Parks and Wildlife Department

In Austin, Hayden Brooks, Molly Stevens, Allen Cooper, and Margaret Russell with the City of Austin Parks and Recreation Department took the lead to form the Children in Nature Collaborative of Austin. In response to a study by Andrew Springer, University of Texas School of Public Health professor of health promotions and behavior sciences, about the relationship between childhood obesity and lack of access to parks where they could engage in unstructured play, Hayden spearheaded funding for an innovative new website called Nature Rocks that allowed parents to easily find parks and connect with outdoor programs in their area.

In the early days, childhood obesity prevention and conservation education were the central focuses of the network. At the time, Texas had three of the five cities with the highest rates of childhood obesity in the country, with only 8 percent of twelfth-grade children in those cities deemed to be physically fit.[20]

In 2012 Nancy Herron was promoted to director of outreach and education at the Texas Parks and Wildlife Department. Her programs included hunter education, boater education, Project Wild, and angler education, which left her little time to focus on Texas Children in Nature activities. Always creative with finding funding within the confines of a state agency, she worked with Anne Brown and Susan Houston at the Texas Parks and Wildlife Foundation to utilize an endowment that had been left to the foundation to fund outdoor classrooms. She found enough funds to create a part-time position to coordinate the effort. Hayden Brooks recalled, "We needed someone who could wake up every day thinking about what was next for TCiN [Texas Children in Nature] and how we could get there at a state level. Nancy made that happen."[21]

After attending the 2010 symposium in my role as a resource manager at Bastrop State Park, I decided that I wanted to focus exclusively on connecting children with nature. While at the park, I worked with naturalist and volunteer Bill Brooks to create a program for a homeschool group to transplant trees from one part of the park to replace loblolly pines dying from the drought. We inducted the children into the "tree army," and they named their trees and geolocated them with a GPS device. Every week I water the trees along with the other rangers. Nine months after we planted the trees, a devastating fire hit Bastrop, destroying more than 80 percent of the park. During the fire, I received a text from one of the rangers, Tony Merino, that simply read, "Jen, we are saving your trees." Those children are now adults, and their trees are still flourishing at the park. The program was a success, but I was ready for more.

In April 2011, just a few months before the fire, I accepted a job with Camp Fire Central Texas as program director and curriculum developer. In that role—and with the support of Executive Director Rebecca Benz (my mentor)—I focused on expanding the reach of the Camp Fire program and its hundred-year traditions to more than thirty after-school programs in the Austin area. Two of the programs I was most proud of helping to develop and execute were the Camp Fire Kids Unplugged and the Camp Fire Family Outdoor Clubs. I also participated in a multitude of Children in Nature Collaborative of Austin gatherings and the early Natural Leaders programs.

In January 2013 I applied to be—and was hired as—the new part-time Texas Children in Nature coordinator (later director). In the first few weeks on the job, I attended a kickoff fundraiser for the new children's museum in San Antonio (later named the DoSeum) with Richard Louv, as well as the environmental literacy symposium with former First Lady Laura Bush. The momentum was building.

In 2014, just before the annual Texas Children in Nature statewide leadership meeting, Jenny Sanders with the Texas Wildlife Association shared that she had taken her son to daycare one day and noticed that the trees around the play area had been cut down. When she asked why, the childcare center director responded, "They drop sticks on the ground, and sticks are dangerous." The next week, her son was crying when she picked him up at the hot South Texas daycare center. When she asked her son what was wrong, he said they had been inside all week because there wasn't any shade on the playground now that the trees were gone.

I shared her story at the meeting. The topic of how to get childcare centers to include more natural spaces and outdoor learning opportunities resonated. Allen Cooper was already working with Robin Moore and Nilda Cosco at the Natural Learning Initiative at North Carolina University to create a book of national guidelines titled *Nature Play and Learning Places: Creating and Managing Places Where Children Engage with Nature*. Allen had the strategic vision to embed those values into the state guidelines for childcare centers; he went to work behind the scenes to make it happen.

In 2015 the national Children and Nature Network selected Texas to host the first International Children and Nature Conference. The conference was held at the Hyatt Lost Pines in Bastrop, Texas, in April, just as the spring wildflowers were at their peak. US Secretary of the Interior Sally Jewell sent a representative to launch the Department of Interior's

Regional leaders being honored at the 2015 Children and Nature Network International Conference, Bastrop, Texas. *Left to right:* Author Richard Louv, Koy Coffer, Colleen McCue, Jaime Gonzalez, Molly Stevens, and Texas Parks and Wildlife Department Executive Director Carter Smith. Photo by Jennifer Bristol

Monarch Awareness campaign, and National Parks Director John Jarvis attended to kick off the Every Kid in a Park initiative.

These major campaigns were inspiring. However, the connections made at the conference had the strongest impact. For example, Stacy McReynolds with the San Antonio Zoo sat next to childhood development specialist Amanda McMickles to listen to renowned landscape architect Rusty Keeler speak about natural play areas. Stacy returned to the San Antonio Zoo with her new connections in hand to further make the case to expand the zoo's preschool programs. Expand they did.

The zoo was able to acquire a four-acre property adjacent the zoo to establish a new preschool focused primarily on outdoor experiential learning. Designed by Lake Flato Architects and incorporating a natural play area designed by New York–based Rusty Keeler, the Will Smith Zoo School opened in 2018. Since its opening, the school has continued to serve as an example for other preschools on how they can say yes to kids picking up sticks, jumping off logs, and playing in the dirt.

Always looking for ways to share the regional best practices, I gave a short presentation in Austin about natural play areas and the benefits of unstructured versus structured play in 2016. After the presentation, a

green school parks. At first, I was resistant to the green school parks project. I thought it was just going to add a sad little garden on the campus that would be forgotten after a few years, like so many others I had observed across the state. But Hayden kept pushing for the element that finally made me a fan of the idea. In addition to creating outdoor classrooms, walking trails, pollinator gardens, and sustainability features, the campuses would invite the community to use the spaces as parks when school was not in session. To me, that was a win, as it increased access to outdoor spaces in neighborhoods that otherwise lacked safe access to parks.

The green school parks project was successful and being considered by other cities. The pandemic made it even more popular, as schools were looking for ways to conduct in-person classes and the outdoors was the safest place. Then, in spring 2022, tragedy struck when a gunman murdered nineteen children and two teachers at Robb Elementary School in Uvalde,

From left: Hayden Brooks, City of Austin Council Member Leslie Pool, and Jennifer Bristol at the celebration of the passage of the Children's Outdoor Bill of Rights and achievements of the Cities Connecting Children to Nature program, January 2017. Courtesy Jennifer Bristol

Texas. After the shooting, the Texas legislature recommended and even funded fences around schools, preventing access to the campuses during in-school and out-of-school time. Despite the change, students still benefit from the natural play areas, outdoor classrooms, walking paths, and gardens. However, in most places the community lost access to those spaces.

The City of Austin embraced the need to continue the Cities Connecting Children to Nature program. They needed an extra funding source to hire a coordinator. Margaret Russell, Molly Stevens, and Nancy Herron (then retired) gathered at my dining room table to frantically hammer out a grant proposal that was due the following day. With the funding secure, the city hired Melody Alcazar to coordinate the effort. Melody had already been a rising star in the movement as a regional leader and champion of nature-based preschools in both the Dallas and San Antonio Zoos. Following the successes in Austin, the cities of San Antonio and Houston were both awarded innovation grants from the National League of Cities.

In 2017, after surveying the network of more than five hundred partner organizations across the state, Texas Children in Nature decided to begin

Margaret Russell (*right*) and Nancy Herron at the Texas Children in Nature conference, 2017. Photo by Jennifer Bristol

hosting annual conferences to share best practices and illuminate the positive work being accomplished. The first conference was held in San Antonio with a theme of nature play and had a waiting list of participants who wanted to attend. The second one was held in Corpus Christi and was equally inspiring.

In 2019 the Texas Parks and Wildlife Department decided to spin off the Texas Children in Nature program to become its own nonprofit. Nancy Herron had retired in 2017, and with her went a strong advocate for the program to remain at the agency. At the same time, Molly Stevens retired from Westcave and Margaret Russell retired from the City of Austin.

I, too, was standing at a crossroads of my career and made the decision to help them launch the nonprofit but was ready to step down as state director. A large part of that decision was my desire to start this book with my mother and a second project with my father. I continued my commitment to connect people with nature through writing and served on the advisory team for the new campsite at Shield Ranch. I also made the commitment to act at a local level by serving as chair of the City of Austin's Environmental Commission and chaired the commission's Bird Friendly Design Committee to reduce bird mortality caused by collisions with buildings. The committee drafted a resolution to embed bird-friendly design into the city's building and design code. The resolution passed in November 2024. Additionally, I served as vice chair for the 2023 Travis County Bond Advisory Committee to put $260 million for new parks and maintenance on the 2023 ballot. The measure represented the most bonds ever put forth for parks by a county in Texas. The measure passed with overwhelming support. In 2024, with the assistance of a board of directors, I started the Victor Emanuel Nature Scholarship Fund to send underserved youth to nature camps that have a focus on bird studies. The fund honors Victor Emanuel's "last big idea" before he passed in 2025.

With a dream team of board members that included Molly Stevens, Nancy Herron, Margaret Russell, Haily Summerford, Hayden Brooks, and Thea Platz, the new Texas Children in Nature Network was formed. Sara Coles was hired as the director.

In 2021, following the success of the Outdoor Learning Environments, a second statewide program developed. Partnering again with the Texas Department of State Health Services, the Texas Parks and Wildlife Department, Texas A&M AgriLife, and the newly formed Texas Children in Nature Network, the program unites with local health agencies to address inequities and increase access to nature. In 2021, fueled by a new grant from the Centers for Disease Control, the Texas Children in Nature Network hired four health liaisons to increase partnerships to provide outdoor programming to underserved urban populations, raise awareness of the benefits of nature, and advocate for more parks and natural spaces.

The legacy of the collaborative effort can be found in every environmental education or outdoor program that was started by one of the partner organizations as an effort to reconnect children with nature. It can be found in the adults who dedicated their careers and volunteer time to create and staff those programs. It lives on in the outdoor classrooms, natural play areas, green school parks, outdoor learning environments, and policy changes needed to move nature out of the "nice-to-have" category and into the "need-to-have" category. Most of all, it lives on in the children who engaged in those programs and connected with the awe and wonder of nature. They are our future stewards.

WOE club members riding at Cibolo Creek Ranch, Marfa, Texas, 2016. Photo by Jennifer Bristol

CHAPTER ELEVEN

THE PRIVATE LANDOWNERS

Texas has one of the fastest-growing populations and has stayed at that breakneck clip for over two decades. In 2025 the population surpassed thirty-one million. What's not increasing is the size of the state. There are approximately 172 million acres of land, with over 95 percent of that being privately owned. Private landowners range from individual homeowners to farmers and ranchers. Churches, private schools, and nonprofits own a surprising inventory of lands as well. And then there are the corporations, both national and international, that own convenience stores, shopping malls, manufacturing facilities, oil refineries, distribution centers, hotels, office parks, and an increasing number of agriculture lands. Each layer has value in our social fabric.

Those private lands do more than just provide a place to live, learn, work, or worship. The vast majority of rain falls on private lands. The rain flows into the two hundred thousand miles of creeks and rivers, more

Teens in the Camp El Ranchito Conservation Corps program out birding, June 17, 2025. Courtesy Jennifer I. Bristol

than one thousand lakes, or ten major aquifers. We depend on those lands to sequester carbon to clean the air and perform a myriad of other ecosystem services. The women in this chapter manage the lands in their care with a generational mindfulness that extends to their families, communities, and state.

SHIELD RANCH: WILD FOR GOOD

Patricia (Pat) Shield Ayres, Vera Ayres Bowen, and Family

In the 1930s Vera Gentry Shield, a former nurse from Abilene, told her wildcatter, oilman husband, Fred Shield, that she wanted a ranch somewhere that would not be covered with messy oil wells. In 1938 the couple purchased the initial acreage of Shield Ranch along Barton Creek in western Travis County. Their home was in San Antonio, where they had access to schools and Fred's work. As ranching operations expanded, the couple raised quarter horses, cattle, angora goats, and peaches.

The couple's only child, Patricia (Pat) Shield, spent summers and school breaks on the ranch. In our 2021 interview, she recounted memories of making peach and wild grape preserves with her mother and riding horses to check on the livestock. She reported that the land looked different then; it was overgrazed, with exposed limestone slopes and

Pat Ayres signing the 1989 conservation easement with The Nature Conservancy at the ranch outside Austin, Texas. Photo courtesy Vera Bowen

pastureland. Her mother loved to fish along the clear, cool waters of Barton Creek.[1]

Vera Shield was fishing at the creek on June 18, 1962, when she suffered a heart attack and died. The *Austin American-Statesman* reported, "Mrs. Shield's dog was standing guard over her body. She was gripping a fishing rod in one hand and a hook she had been baiting in the other. . . . The dog did not want the rescue party to touch the body of its 57-year-old mistress."[2]

At the time of her mother's passing, Pat had already graduated from Wellesley College in Massachusetts and married Robert M. Ayres, who also grew up in San Antonio. The loss drew Pat into taking a stronger interest in the management of the family properties, even though she was raising two young children, Bob and Vera. Thankfully, her husband shared a passion for the land and ranching.

While living in San Antonio and raising their children, Pat became involved with social justice issues, especially pertaining to youth, and worked with Good Samaritan Community Services, which was part of Saint Mark's Episcopal Church in San Antonio. In 1971, during a time of reform, Governor Dolph Briscoe appointed Pat as the first woman to serve on the Texas Youth Council. In 1975, just after becoming chair of the commission, Pat led a charge to develop and adopt a bill of rights to protect youth within the juvenile detention system.

Following the story, *The Marshall News Messenger* ran two stories in December 1975. The first applauded the reforms and bill of rights.[3] The second highlighted the toxic culture Pat had to overcome to be an effective leader. "After only a week in office, Mrs. Patsy Ayres of San Antonio has been stripped of her vote on the Texas Youth Council. The move to prohibit Mrs. Ayres from voting, except to break a tie, was led by Don Workman."[4]

Pat persisted, and a judge ruled that the motion to block her from voting was illegal and must be reinstated. Under Pat's leadership, the council pushed for reform to decentralize state schools and instead advocated for community retention centers and programs to move youth out of the criminal justice system back into society. The hope was to keep youth from unnecessarily being locked up rather than being offered pathways to a stable life outside the system. The state school in Crockett even added a wilderness camping program as part of its services to rehabilitate neglected and abandoned youth.[5]

Vera Ayres Bowen, Pat and Robert's daughter, recalled that period of time as being very busy for her mother, who constantly was on the move between the ranch, meetings in Austin, and other areas of the state. "I went with her to a lot of the schools and halfway houses. She regularly wanted to check on their progress, welfare of the kids, and programs. She traveled so much that she kept a CB radio in her wood-paneled station wagon so we could stay in touch with her. This was long before cell phones."[6]

Wanting to use their money for good, Pat and her husband formed the Shield-Ayres Foundation in 1977 as a mechanism to support projects that aligned with their values, including a large number of conservation programs. Upon graduating from college, their son, Bob Ayres, and his wife, Margie, became trustees. Bob and Margie lived in California for a time, but they returned to Texas in the 1980s and subsequently welcomed two children, Kathleen and Elizabeth, into their family. Bob then assumed an active role managing the family's properties. Vera, who had been living in Germany with her husband, moved back to Texas in 1989 to raise their four children, Marshall, Gentry, Sophie, and Mallory. At that time Vera also became a trustee of the foundation. Their return was fortuitous, as the ranch was about to face its first major threat.

Pat Shield Ayres (*right*) and Vera Ayres Bowen at the ranch, 2019. Courtesy Vera Bowen

In the late 1960s a plan was proposed to construct a highway, called the Outer Loop, to encircle Austin with eighty-plus miles of roadway. The planned routes for the loop varied over the next few decades, and for a time, the project was shelved. Austin experienced rapid growth in the early 1980s, which fanned the flames of the project, to become a raging fire by the mid-1980s. The idea of the Outer Loop sent land speculators racing for the hills along Ranch Road 2244 and Highway 71 to buy every property they could.

With speculators eyeing their property, the Ayres family came together to discuss the future of the ranch. Every idea was open to consideration: sell it as a whole, break it up, keep it intact, or divide it among each family member. Collectively, they agreed to resist the temptation of making a quick buck and decided to keep the property intact, subsequently developing the ranch's first master plan to guide its management. From that point forward, Bob immersed himself in learning about best practices to manage the lands holistically. For his efforts, he emerged as a trusted leader in the conservation community who often bridged the gap between private landowners and conservationists.

The sixty-eight-hundred-acre ranch is home to endangered Golden-cheeked Warblers and threatened Black-capped Vireos, and it protects one of Austin's most valuable water sources, Barton Creek. According to the ranch's master plan, "The Shield Ranch is bisected by the 98th meridian, often considered the gateway to the American West. The ranch lies within the Balcones Canyonlands of the Edwards Plateau and the Barton Creek watershed. This watershed drains to the Edwards Aquifer recharge zone, which is impacted by the hydrologic performance of Shield Ranch and neighboring, large-scale land holdings. The region is considered a 'danger zone' due to anticipated impacts of climate change paired with rapid population growth."[7]

In 1998 the family placed 4,670 acres of the ranch into a conservation easement with The Nature Conservancy. In accordance with the master plan, a second easement "was sold to the City of Austin in 1999 as part of a voter-approved initiative to protect water quality in the Barton Springs Segment of the Edwards Aquifer."[8] The conservation easements protect more than 90 percent of the ranch from development, while still allowing for a family residence and ranching, hunting, and recreational activities. In 2000 the careful management of the land was recognized with the Texas Parks and Wildlife Department's Lone Star Land Steward Award.[9]

Vera shared that, while she felt connected to the ranch and was grateful her children could swim in the same swimming holes as she and her mother did, something was missing. The land kept singing to her in a low, gentle voice; however, she could not truly hear the song until it included the voices and laughter of a multitude of children.

In 2007, in collaboration with Westcave Outdoor Learning Center and El Buen Samaritano Episcopal Mission, the family allowed a portion of the land to be used as a nature immersion summer camp for underserved children. That new purpose for the land was the turning point for Vera; she was all in.

The camp became one of the community-based solutions that the Texas Children in Nature Network and the local Children in Nature Collaborative of Austin were calling for. Under the leadership of Board Chair Hayden Brooks and Executive Director Molly Stevens, Westcave Outdoor Learning Center began exploring various collaborations to connect children in underserved communities who had limited or no access to outdoor opportunities. The partnership with Shield Ranch seemed like the perfect bridge between a public need and a private landowner. It was something that was not yet being done.

The family saw the value in the program and, in 2008, established the Shield Ranch Foundation to operate Camp El Ranchito on the ranch. Vera increasingly took the lead in supporting each aspect of the camp as it grew. Naturalist and Camp El Ranchito founding director Terri Siegenthaler designed the core curriculum and activities.

The majority of children attending had never had the opportunity to camp outdoors before. For the first fourteen years, campers slept in tents that they pitched and dined in a pop-up screened dining hall. On more than one occasion, campers and staff had to evacuate quickly to a nearby church as severe weather erupted. Despite floods, droughts, and heat, the children continued to return year after year.[10]

In 2016 the family convened for another visioning session that included all layers of the family from Pat to the grandchildren. During the process, the family decided it was time Camp El Ranchito became a permanent

families, engage in team building activities, and connect with the natural world. Many participants had not been camping before and were in awe of the dark skies at night and the ability to live sustainably off the grid at the campsite. The loop trail hike we did was a big peak accomplishment."[12]

The legacies of Pat Ayres and Vera Bowen and their families continue to grow and mature. As Vera asserted, the ranch and the land were always special places that she valued and wanted her own children to share. However, when the camp idea emerged, it became the focus and purpose she didn't know she needed, inspiring an even deeper and more spiritual commitment to the land. Each camper leaves with an ethos of stewardship and love of nature, carrying that forward into their own communities.

When the family said they had an "abiding love of nature so deep that it had to be shared,"[13] they meant it. The family's successful effort to open their land in such a way is a remarkable achievement, involving years of finalizing mundane details such as liability, infrastructure improvements, and maintenance—and a plan to fund it all. Thanks to the efforts and actions of Pat, Vera, and Bob (in partnership with the rest of the family) and fortification from the conservation easements' restrictions, the family was able to persevere and resist the temptation to throw in the towel and sell the land for development. The family has placed their additional properties in Texas under conservation easements to protect water systems, and a myriad of endangered species and a growing list of species of concern.

Thanks to the vision and care of Pat, Vera, Bob, and their families, Shield Ranch and Camp El Ranchito will always be places where children can play, explore, and fall in love with nature—and not just the nature of some faraway place they see only once a year; it is the nature that they are surrounded by in their daily lives. The family's legacy most assuredly continues in their children and a growing number of grandchildren, who carry the torch to be "wild for good."

PRIVATE LANDS WITH A PUBLIC PURPOSE

Paula Young Smith

Paula Smith was born and raised in Denton, Texas, but admits that she did not develop a strong connection with the outdoors as a youth. The daughter of a professor, she was more interested in books and her friends than stopping to notice nature.

As a newly hatched lawyer fresh out of the University of Texas Law School, Paula made her first foray into advocating for nature by trying to help protect trees on the university's campus. In 1969 the University of Texas was expanding its football stadium and parking area by several acres, with little regard for the centuries-old live oaks along the slopes of Waller Creek. Students were rallying to the cause and holding protests to "save the trees."[14] One report states, "In an attempt to stop the bulldozing, student protesters chained themselves to trees, and [Frank] Erwin [chair of the University of Texas System Board of Regents], complete with hard hat and bullhorn, personally oversaw their arrests."[15]

Paula and a group of other attorneys filed an injunction to stop the bulldozers from harming the trees. Unfortunately, the bulldozers took only a three- or four-day pause before firing up their engines again. Paula shared that Frank Erwin appeared at the construction site and demanded that the largest trees be cut down first.[16] It was not a high point for the university's long and storied history or for Frank Erwin.

It also was not a catalyst that propelled Paula into a life of environmental activism, nor did it stir a desire to chain herself to trees. She was busy raising two young boys, Brian and Carter, with her husband, Ernest Smith, and balancing home life with the demands of being an attorney. Ernest was a renowned oil and gas professor, who subsequently became dean of the Law School at the University of Texas–Austin.

In 1982 Ann Richards was elected as the first woman state treasurer in Texas. Ann asked her friend Paula to head the Unclaimed Property

Paula Smith placing a monitoring tag on a monarch butterfly at Dobbs Run Ranch, October 2012. Courtesy Paula Smith

Program. The program was wildly successful; between 1983 and 1990 it reunited taxpayers and businesses with more than $288 million in unclaimed property. What was not claimed eventually was transferred to the state. The program even reunited then–Texas Rangers managing partner George W. Bush with $12,000.[17] Paula and Ann received high praise for the program in a *Fort Worth Star-Telegram* article: "Ann Richards and Paula Smith, who heads the Unclaimed Property Division, are sort of ahead of their time."[18] Ann Richards credits Paula's work in the division as one of the main reasons she was able to win the election to become the second woman governor of Texas.

Seeing an opportunity to assist other states and companies in understanding the laws surrounding unclaimed property, Paula started a business to provide reporting services to financial institutions and nationally based corporations. After the business sold to State Street Bank, she was obliged to commute between her home in Austin and her work in Boston.

In 1996, Paula, Ernest, and their conservation-minded son Carter decided to purchase a ranch after Ernest sold his interest in another ranch that he owned with other family members. The three made a list of the amenities they wanted on the property: Ernest wanted solitude, Carter was interested in good habitat for hunting, and Paula wanted a source of flowing water outdoors—and indoor plumbing.[19]

What they found was an old five-thousand-acre goat ranch located along the Hill Country convergence zone of the Edwards Plateau, Chihuahuan Desert, and Trans-Pecos ecoregions between Rocksprings and Bracketville, Texas. Paula identified acquisition of the ranch as the catalyst that finally fanned the flames of her interest in nature. "I am not a self-taught naturalist. Once I became interested, I had a lot of people teach me about it. The more I learned, the more I wanted to know," she shared in a 2022 interview.[20] She and Ernest both became Texas Master Naturalists to better understand the plants and wildlife that call the ranch home.

The west prong of the Nueces River flows intermittently through the property and rises and dries with the fickle rains of West Texas. Despite the river being a wet-weather waterway, spring-fed pools remain even in harsh droughts. The Smiths began a multiyear Ashe juniper control plan to thin the water-loving tree species. Since the Ashe juniper is also a critical component of endangered Golden-cheeked Warbler and threatened Black-capped Vireo habitat, the Smiths are careful to leave stands of mature junipers to support the birds that arrive in March to nest and breed.

The family also manages the habitats of two other endangered species: the Texas snowbell and Tobusch fishhook cactus. Many landowners in Texas find having endangered species on their property to be a nuisance and even go to great lengths to destroy the habitat that supports them. For Paula, caring for these species brings deep joy and purpose. In 2000 her son Carter, then state director of The Nature Conservancy of Texas, recommended that the family place much of the ranch in a conservation easement to offer long-term protection for the land, water, plants, and wildlife.

Artist and writer Margie Crisp describes in her book *The Nueces River: Río Escondido* spending an afternoon with Paula: "The best part of being out on the ranch with Paula Smith is her unrelenting curiosity about the natural world and her infectious passion for Dobbs Run. . . . The Smiths have worked to create a mosaic of grasslands and woodlands in the canyons and on the hills. The goal, Paula tells me, is to support and encourage diversity of plants and wildlife by developing niche habitats."[21]

To understand those niche habitats, Paula and her family have allowed their property to be used as a laboratory of sorts, where species can be studied by university students and Texas Parks and Wildlife Department biologists. For example, a graduate student from Texas A&M University–Kingsville conducted studies on Montezuma Quail. The elusive quail, once abundant, now exists in dwindling numbers. The research has helped biologists understand the ground-dwelling bird's preferred habitat and behavior so other landowners can manage their properties favorably for the species. The ranch has also welcomed multiyear studies on the Golden-cheeked Warbler and Black-capped Vireo. With 95 percent of Texas lands privately owned, it is primarily up to families and corporations to manage for species of concern.

In addition to Golden-cheeked Warblers, Black-capped Vireos, and Montezuma Quails, Paula wanted to improve raptor populations. She shared that she was able to release two rehabilitated Bald Eagles at the ranch on her birthday. "It was the best birthday gift ever. Releasing them and watching them soar."[22] Eagles once were plentiful in the region, but years of ranchers hunting and poisoning them has made their recovery from near extinction difficult in West Texas.

Releasing two rehabilitated Bald Eagles with Last Chance Forever rehabilitators John Karger and Melissa Hill, Judge John Rainey, Cynthia and Jim Burgin, Paula Smith, Ernest Smith, Brian Smith (son), and other friends, March 17, 2019. Courtesy Paula Smith

Each May, Paula and Ernest mobilize several of their Texas Master Naturalist friends and interested biologists to conduct a butterfly count as part of the North American Butterfly Census. On occasion, during the fall migration, they net monarchs to place tiny tags on their wings to help scientists understand populations and migration routes. In 2022 the iconic monarch butterfly was placed on the international endangered species list. Restoring native plant communities that bloom in the spring and fall as the butterflies migrate is critical for fueling them along the way. Paula takes great pride in the variety of native flowering plants, including native milkweed, they have restored and encouraged on the once overgrazed landscape.[23]

One of those flowering plants, the Texas snowbell, also just happens to be an endangered species. The flowering bush is native to the drainages and canyons of the Nueces and Devils Rivers. The lush plants are favorites of native bees and butterflies; however, they also are favorite snacks for whitetail deer and goats. After purchasing the property, the Smiths enlisted botanist Jackie Poole of the Texas Parks and Wildlife Department to conduct a plant survey. Jackie initially was looking for endangered Tobusch fishhook cactus but began to suspect that Texas snowbell might be on the property as well. Paula described Jackie as being so skilled that, after hours of searching the canyons and cliffs with Ernest, she found a single petal from the Texas snowbell and knew there must be a few remaining bushes high on the cliffs, where they are protected from deer and invasive aoudad goats.

Friend and fellow landowner J. David Bamburger heard about the presence of the snowbells at Dobbs Run Ranch and embarked on a mission to do what he could to preserve them. Paula, Ernest, Carter, and J. David convened a meeting with nineteen local ranchers in the Nueces watershed to enlist them in a long-term conservation effort. Paula remembered,

Ernest Smith, J. David Bamburger, and Joanna Bamburger with Paula Smith and friend at one of the snowbell recovery sites, 2014. Courtesy Paula Smith

The Texas Youth Hunting Program at Dobbs Run Ranch, 2012. Courtesy Texas Parks and Wildlife Department

"J. David came to collect seeds from our snowbells to cultivate. Once the plants were large enough, we planted them with a protective wire cage surrounding them so the browsing wildlife couldn't get to them. J. David raised saplings from the seeds collected on Dobbs Run at his ranch. He came out to plant the first ones on our property with us. Over the course of a decade, we planted over a thousand plants with J. David. We were pleased that several of the adjacent landowners wanted to participate in the recovery project. It takes that type of cooperation to successfully do something like this."[24]

One element of the success of caring for the land to support endangered species and ensure healthy habitats for all species is managing grazers and rooters such as white-tailed deer, feral hogs, and other exotic game species. The Smiths have a careful management plan that guides them on how many deer or turkey can be taken in a season. Since 2012 the family has opened the ranch to the Texas Youth Hunting Program. In a 2013 article in *Texas Parks and Wildlife Magazine*, Carter explained the overall purpose and structure of the program: "TYHP [the Texas Youth Hunting Program], which was created as a partnership between the Texas Wildlife Association and the Texas Parks and Wildlife Department, facilitates mentored parent-child hunts for new and beginning youth hunters on private lands. The goal is to expose kids in a safe environment to all elements of hunting—from 'field to fork,' as one TYHP leader described it. Now in its seventeenth year, TYHP pairs these hunters with willing landowners and teams of trained volunteer 'huntmasters,' who lead the youth hunts and serve as instructors, mentors, safety officers, guides, and camp cooks. To date, thousands of kids, more than half of whom had never hunted before, have participated in the program."[25]

More than eighty kids and their accompanying adults (mom, dad, aunt, uncle, or grandparent) have participated in the program at Dobbs Run Ranch. Hunts are held during deer season; however, youth and adults can harvest a feral hog or exotic species in addition to their limits of white-tailed deer. Learning to hunt gives kids a sense of self-reliance and independence that builds their core confidence, while also teaching them about the value of a life. The weekend includes talks by state game wardens and information about native plants and wildlife habitat. For many kids, it is also the first time they experience camping, outdoor cooking, and stargazing away from an urban center.

Always looking for ways the ranch can be used for education and study, the Smiths partnered on occasion with the Texas Wildlife Association to host every fourth grader from Bracketville Independent School District to offer them hands-on, in-the-field lessons about land, water, wildlife, and stewardship. As part of that partnership, they hosted a teacher training program to study the ecosystems of the region.

In addition to offering the ranch to researchers, Texas Master Naturalists, youth hunts, and students, Paula and her family include the property as part of their philanthropy. The Smith family has donated guided bird tours (with all the bells and whistles) to Travis Audubon, Audubon Texas, and Devil's River Conservancy fundraisers. Ernest served for many years on the board for the Appalachian Mountain Club, and one year, he provided the ranch's facilities and miles of hiking as a donation for an event benefiting its members.[26]

Since 2018 they have opened their ranch to the Texas A&M AgriLife Extension Birding on the Border festival. Following the success of the Birding on the Boarder program, Texas A&M AgriLife Extension approached Paula about hosting campers from the Rio Diablo Birding Camp. Since 2021 the ranch has welcomed more than fifty young birders and their guides to the ranch to explore the canyons, ridges, and upland regions of the ranch.

The ecosystem services the ranch provides extend well beyond its borders: habitat for endangered migratory species, carbon sequestration, and upstream water conservation, just to name a few. The ranch resides within the Edwards-Trinity Aquifer Contributing Zone, which means every drop of rain that lands on the ranch has the potential to become drinking water for someone drawing from the aquifer. In 2023 the Smiths purchased an adjacent property to protect Cherry Creek Springs that flows into the Nueces River. That purchase expanded the ranch to 9,100 acres.

Paula's conservation legacy lives on through the love of land and nature exhibited by her sons Brian and Carter, along with Carter's wife, Stacy, and their son, Ryland. "Mom treasures few things more than

Paula Smith with her grandson, Ryland Smith, on the porch at Dobbs Run Ranch, November 2021. Courtesy Carter Smith

sharing her love of nature at the ranch with friends, family, and even perfect strangers," said Carter. "Even the most cryptic wildflower and the smallest butterfly gives her great joy and sparks her unending curiosity. I must say, her enthusiasm is infectious, and people always leave the ranch a lot more enlightened about the natural features and plants and animals

Campers and guides of Rio Diablo Birding Camp at Dobbs Run Ranch, June 13, 2025. Photo by Jennifer L. Bristol

of the Hill Country after spending time with her trekking over the rocks and hills of Dobbs Run. It is fun to behold."[27]

Although she could not save the oak trees from the bulldozer on the University of Texas campus, Paula has been able to save the trees, birds, butterflies, bees, flowers, and other creatures of Dobbs Run Ranch forevermore.

system of water storage tanks (otherwise known as ponds outside Texas) to capture the limited annual ten to eleven inches of rain. Each tank was better than the last, until they finally constructed a large, nine-foot-deep pool to harness the water. Carolyn and Sherwood enjoyed growing things and soon planted multiple trees around the tanks. The reintroduction of water and trees along the arroyo created an oasis for migrant and resident birds.

Carolyn's sister, Andrea Ohl, liked what Carolyn had built on her property so much that she bought an adjacent tract of land just down the hill, where she constructed her own earth-shelter home. Andrea, a noted archaeologist and writer, joined the Center for Big Bend Studies at Sul Ross State University in 1997. The two industrious sisters continued to document the geology, archaeology, and biodiversity of their slice of the Christmas Mountains. In the early 2000s Carolyn, by now in her sixties, converted her journal into an online blog, recording her observations and the progress with her oasis project. She began incorporating photographs into the blog, which captured the attention of many a birder and sparked their interest in visiting the region and her oasis.

Carolyn lost her beloved Sherwood in 2003. Never one to stay idle, she married Hugh Johnson a few years later. By the mid-2000s, as she continued to cultivate habitat to support more than 275 species of birds, Carolyn was welcoming more and more birders to her property. She was careful to plant a mix of taller, native trees; understory shrubs; and grassland plants to feed the birds year-round. The flowering plants attracted a wide variety of hummingbirds, which she began to support further with feeders. She also installed several shallow water features where small birds could bathe and drink. The water features and feeding stations are surrounded by observation areas where birders can sit quietly to view the wonders of the oasis.

The observation area is not fancy, with its aging office chairs and donated barstools, but what it lacks in design it makes up for in function with shade to ward off the ever-present sun. In a 2011 interview with noted birder and journalist Gary Clark, Carolyn shared this about her oasis: "I think life's meaning is to create and share. . . . When people visit and enjoy their first Lucifer [Hummingbird], I enjoy it all over again vicariously through them."[30] The oasis is one of the few places in the

United States where the Lucifer Hummingbird can reliably be observed when in Texas.

In a *Passport to Texas* radio interview with Cecilia Nasti of the Texas Parks and Wildlife Department, Carolyn shared, "I can be sitting here, just looking at the same old stuff, and bet money that nothing interesting's gonna come along. And there, all of a sudden, oh my gosh, there's a lifer! But it won't happen if I'm not sitting here looking."[31]

Life-list birds abound at the oasis for many birders wishing to find certain species in the vast wilds of the Big Bend region. The beauty of Carolyn's efforts is that she makes it relatively easy for advanced and intermediate birders alike to observe species such as the Varied Thrush, Elf Owl, Golden Eagle, Chihuahuan Meadowlark, and more than thirty species of warblers. But birds are not all that flock to the flowering plants, shade trees, and life-giving water. Carolyn has documented an entire parade of butterflies and moths as well. Occasionally, she has to fight off larger species such as peccary, deer, and even a black bear that was interested in the water and food source at the bird feeders. Of course, all the abundant animal activity is captured on her blog as she makes observations and photographs the events.

After remarrying, Carolyn moved into Alpine and began transforming her home there into another bird, butterfly, and bee paradise, complete with a small wetland, flowering plants, trees, brush piles, and feeding station. Always wanting to share the glory of nature with others, she continues to welcome visitors, asking them to go birding with her or just sit in the shade and observe the winged wonders dancing about. Despite living in Alpine, she makes the trek to her oasis in the mountains most every day—a three-hour round trip.

In 2021, with support from Cecilia Riley, Madge Lindsay, and board members of Trans-Pecos Bird Conservation, the oasis received a $5,000 conservation grant from the Texas Parks and Wildlife Department's Greater Texas Birding Classic program. Entrance fees paid into the birding classic support grants for which nonprofits can apply to restore habitat, enhance wildlife viewing projects, or create educational programs. The grant helped pay for a new rain-harvesting tank and greatly improved the viewing area. (The office chairs remained.) With more than ten species of hummingbirds being observed at the oasis, including the Blue-throated

At the Christmas Mountains Oasis on the Great Texas Birding Classic, 2022. *Front row, from left:* Madge Lindsay, Beth McBroom, Carolyn Ohl, Cecilia Riley, and Linda Hedges. *Second row, from left:* Michael Gray, Bill Sain, and Rick Reese. Courtesy Madge Lindsay

Mountain-gem, the oasis is an important stop in the Davis Mountains Hummingbird Festival, which is also managed by Cecilia Riley, Madge Lindsay, Linda Hedges (former wildlife biologist with the Texas Parks and Wildlife Department), and Trans-Pecos Bird Conservation.

I have had the privilege to visit the oasis several times and will note that, like a hummingbird, Carolyn rarely sat still during those visits. On one occasion, I was with my husband and mother on the Great Texas Birding Classic. When Carolyn saw my husband, she quickly asked for his help unloading a new water pump from the back of her truck, which he was happy to oblige. On our visit in the spring of 2022, Carolyn shared how nervous she was about the lack of rain. "If we don't get some good rains this season, I'm going to have to buy water again, and that is getting expensive. I'm not sure I can continue if we don't get some rain."[32]

The rains did return in June 2022, and the oasis got a much-needed drink. That same month, however, Carolyn was dealt a double blow when her daughter had to have a brain tumor removed and her husband was airlifted to the intensive care unit in Odessa, Texas. As if that were not enough, the tank she had stuccoed by hand in the 1990s started to leak significantly. Her son stepped in to help with the maintenance of the oasis. Considering Carolyn turned eighty-two in 2022, that was good timing. It seems Carolyn does not sit still to age either.

Carolyn is working with her son and local conservation groups to ensure that the almost sixty years of moving rocks, building tanks, patching leaks, planting trees and shrubs, clearing brush, feeding birds, and welcoming guests to her oasis do not fade like the desert flowers. She is a great example of how someone with limited means can enhance their property through native plants and a water source to make it a more supportive habitat for wildlife. While six hundred acres might seem like a vast empire, it is actually a relatively small piece of property in the hardpan land of ranches that reach into the hundreds of thousands of acres in the western part of Texas.

With the exception of the Texas Parks and Wildlife Department's grant and small donations from visitors, including a GoFundMe project to purchase a sixty-five-thousand-gallon metal storage tank, Carolyn has self-funded all of her projects. Because of her willingness to share her property and observations on her blog and social media, she has inspired countless others to visit the region and learn more about the fragile ecosystem of the Trans-Pecos / Big Bend area.

the most part, digital natives who are comfortable using social media and new methods to reach multiple audiences and amplify the issues they care about. These women are moving beyond the time when race, religion, gender, or physical differences divided us to boldly taking us into the future.

Children learning about bird adaptations from state park interpreter at Estero Llano Grande State Park, 2013. Courtesy Jennifer Bristol

HELPING BLACK BIRDERS SOAR

Danielle Elizabeth Belleny

Born in Dayton, Ohio, Danielle Belleny moved with her family to San Antonio when she was three years old. She reports having had a strong connection with nature, especially animals, even as a youth. She set out to earn a degree in veterinary science from Texas A&M University in College Station but later switched to wildlife management at Texas A&M University–Kingsville.

Danielle Belleny ready to bird, *Texas Parks and Wildlife Magazine* Trail Blazer issue, April 2022. Courtesy Texas Parks and Wildlife Department

While living and attending college in South Texas, Danielle "got hooked on birding," as she puts it. The diversity of birds in her area "grabbed her attention, and she realized she was on the right path."[1] After graduating from Tarleton State University with a master's degree, she joined Plateau Land and Wildlife Management in Austin, Texas. While birding together in 2021, Danielle shared with me that it was not always easy being a young African American woman pursuing a wildlife management degree. Sometimes, she was one of only a few women—and often the only person of color—in her classes.[2]

May 2020 was an explosive month in the nation's history. On

May 25, 2020, an African American man named George Floyd was violently murdered by a white police officer. Just days before, an African American man named Christian Cooper encountered a racist white woman while birding in Central Park in New York City. Both incidents were caught on video, which flooded the media and fueled passionate protests across the country.

Danielle joined a video conference call with a group of other scientists, biologists, and educators, who decided to take positive and immediate action. They created Black Birders Week and launched it on May 31, 2020. The National Audubon Society and Cornell Lab of Ornithology liked the concept and helped with financial and media support. Danielle managed the group's social media efforts for the first year and was part of the forty-plus-person leadership team. The team and program are now part of a larger collective called Black AF in STEM, which "seeks to support, uplift, and amplify Black STEM professionals in natural resources and the environment through professional development, career connection, and community engagement."[3] Black Birders Week continues to be a popular annual event each May.

Using her voice to engage people of color with the outdoors, Danielle has contributed multiple articles and joined several podcasts of the American Birding Association. In 2022 she published her first book, *This Is a Book for People Who Love Birds*. Her second book, *Junior Birder's Handbook*, followed a year later. Both feature beautiful illustrations. All of her published works and social media engagements invite the reader or listener in a welcoming and inclusive way to explore in nature.

Danielle's contributions to conservation will no doubt continue to grow as she follows her passions and inspires young women and men of color to pursue science-based careers and adventures in the outdoors.

OPENING DOORS FOR EVERYBODY—AND EVERY BODY—THROUGH BIRDING

Virginia Rose

In 1973, when she was fourteen, Virginia Rose was riding her dapple-gray Arabian horse when it spooked and threw her into a fence. The injury left her paralyzed. Ever since, she has moved about the world in a manual wheelchair. She certainly had challenges, but she overcame them with her indomitable spirit to ensure a rich and purpose-filled life.

Virginia taught college and high school English until she retired in 2016. In the early 2000s, after a long day at school, she decided to attend a Travis Audubon lecture on the habits of House Finches. The topic delighted her, but more importantly, she instantly connected with the other people in the room. In that moment, she began her birding journey. She took classes and participated in field trips with Audubon, even taking bird-oriented vacations. In an interview with *BirdCallsRadio*, Virginia shared, "Travis Audubon teachers and field trip leaders never said no to me, not once. They took me everywhere the other walking birders went."[4]

In 2010 she began leading Beginning Birding events with Travis Audubon, and in 2017 she joined its board of directors. Over the course of fifteen or so years, birding awakened something in her. She poetically describes the transformative awe and wonder of nature: "Sitting entirely alone, tucked tightly into a springtime, aromatic East Texas forest, I was spellbound by birdsong. The ethereal fluted tones of [spring migrants] were everywhere, and I was stunned to realize I was meeting my best self, happier than I'd ever been."[5] Virginia wanted other people with mobility challenges to experience the joys she had discovered observing birds and exploring nature.

A watershed moment occurred in 2018 when she decided to host a Birdathon challenge with Travis Audubon. Birdathon is an event where teams birdwatch all day to help raise money for the organization. She named her team Birdability and focused on birding at five accessible

A RISING STAR IN THE RIO GRANDE VALLEY

Marisa Oliva-Rodriguez

Born in San Antonio, Texas, in 1974, Marisa Oliva-Rodriguez described her childhood as full of outdoor adventure. "My parents would pack us into the car, and we would end up in the Hill Country, walking through fields of bluebonnets or end up in a park somewhere. It was a big influence to have those adventures."[8]

In high school, Marisa joined a High Adventure Explorers Scout group to participate in hiking, camping, canoeing, and other outdoor pursuits. She loved the program but couldn't help but notice, "I never ran into anyone who looked like me. There were never African American or Hispanic

Marisa with her family, Julia, Alex, and Roy Rodriguez, December 2022. Courtesy Marisa Oliva-Rodriguez

people out on the trails. After college, I pursued a career as a wildlife and fisheries biologist. I also began volunteering with Girl Scouts and other groups to teach them about the environment and outdoors."[9]

Marisa attended Texas A&M University, earning a bachelor's degree in wildlife and fisheries science, and later earned her master's in forest resources and wildlife ecology from Pennsylvania State University. Not long after graduating, she moved to the Rio Grande Valley to work as a park ranger at Bentsen Rio Grande State Park. In 2004 she became the first Latina manager of Edinburg Scenic Wetlands and World Birding Center. The forty-acre Edinburg Scenic Wetlands was the first of the World Birding Center sites to fully open in 2003. As manager, Marisa used the center as a place to develop programs to engage the community, with a strong focus on families and kids.[10]

Marisa listened to families to help her understand the barriers that prevented them from participating in outdoor events. They highlighted several impediments, including a lack of evening or after-work programs that were low cost or free. Hearing that, she worked to secure grants and funding to provide nature programs, including the popular Kids Birding Clubs. She also recognized that transportation was an obstacle and secured a van to shuttle kids and volunteers to other nature centers and events.[11]

In our interview, Marisa shared that when she first arrived in the Rio Grande Valley, there were very few Hispanics leading bird and nature walks or interpretive programs. She wanted to change that so the kids coming through the program could engage with rangers, biologists, and interpreters who looked and spoke like them. Over the years, she trained dozens of young Hispanic and Latinx men and women, who later went to work at other parks and nature centers both in the valley and across the state.

Collaboration is something Marisa values, so she endeavored to bring the valley's nature centers together to improve communication and partner for valley-wide events. She also became involved with the Association of Nature Center Administrators, a thirty-year-old organization that strives to advance the significance of nature centers across the country. Marisa and others in the valley—and across Texas—wanted to showcase the network of twenty-plus nature centers and birding centers located in the Rio Grande Valley. Marisa and Colleen Hook at the Quinta Mazatlan World Birding Center in McAllen spearheaded the Dream Big Summit, bringing world-class training to the nature centers and creating networking opportunities.[12]

Marisa, a mother of two, wanted to be more involved in connecting children with nature and, in 2016, became the Texas Children in Nature Network regional leader for the valley. She also was part of a pilot program that focused on creating outdoor learning environments at daycare centers. Part of that process included training childcare workers on how to deliver lessons and devise activities to facilitate nature play. Marisa shared that she was proud to train more than two hundred childcare workers in the valley on the Growing Up WILD curriculum. She also encouraged childcare centers to redesign their outdoor spaces to foster more active play and a deeper connection with nature. In 2022 Marisa stepped down as the manager of the Edinburg Scenic Wetlands and World Birding Center and joined the Texas Children in Nature Network's team as a health and nature liaison.

When asked to share what social or cultural barriers she encountered in her career, Marisa stated, "Being Hispanic and a woman, my words and opinions have often been dismissed. I'm asked to be at the table, but in the past especially, there are people that don't really want me or other women to say anything. It's aggravating. But I just keep on trucking and it is changing and getting better as more Latinas get involved. I've also always had a great network of support to help me get past those times that my opinion was dismissed."[13]

Marisa's career and commitment to conservation continue to grow and unfold, leaving her legacy only halfway realized. However, the thousands of visitors to the Edinburg Scenic Wetlands and World Birding Center will continue to benefit from her vision that all people deserve a safe and inclusive space to enjoy nature. That vision also lives on in the dozens of Hispanic and female staff she hired and supported who might not have otherwise had those opportunities. In her new role with the Texas Children in Nature Network, she will continue to utilize her leadership skills to collaborate across the valley to bring together the conservation, health, education, and built environment communities to find more ways to connect children with nature.

RECONNECTING THE AMERICAN BISON AND THE LIPAN APACHE

Lucille Contreras

For centuries, American bison grazed the abundant grasses and plants of the prairies from Alaska to northern Mexico. As the bison wandered and moved within their range, they crushed the hard ground with their feet, pressed seeds into the fertile earth, or transported seeds that clung to their thick, curly hair; their manure, urine, and bones fertilized the soil.

The largest terrestrial animal in the Americas provided the Indigenous people of North America with food, shelter, and clothing. The American bison is known by many names; for the Lipan Apache, they are *Iyanee'*. The greater Apache nation once covered much of the Southwest, spanning segments of Mexico, Arizona, New Mexico, Colorado, and

Lucille Contreras with the bison at her ranch near Waelder, Texas, 2022. Courtesy Lucille Contreras

Texas. The eastern community that lived in what is now West and South Texas was known as the Lipan Apache.

After the Civil War, Texas and the western states and territories continued the genocide that the United States had perpetrated against Indigenous people in the East. The Apache tried to hold fast to their homelands, but in the end, they were no match for the military, well-armed settlers, and waves of European diseases. In the 1880s all remaining Lipan Apache were removed from Texas to reservations in Arizona and New Mexico.

At the same time, US policies called for the destruction of the bison. Millions of bison were slaughtered by "buffalo hunters" and left to rot where they fell. The genocide of the people and the species, coupled with destructive agricultural practices, led to a radical change in the prairies' ecosystems. Thankfully, Mary Ann Goodnight, wife of legendary cattleman Charles Goodnight, took a passionate interest in the bison. Of the thirty million to forty million bison that once roamed the plains, only around three hundred remained at the close of the 1800s. Mary Ann rescued two bison calves and from that point on, she and her husband began cultivating a small herd. Over the years, they shared their stock with zoos and other bison conservation groups to preserve the species. The descendants of those calves Mary Ann rescued live on today in the state herd at Cap Rock Canyon State Park and the former home of the Goodnights at Palo Duro Canyon State Park.

Since Lucille Contreras was a young girl growing up in urban San Antonio, she always dreamed of living in the country. The Lipan Apache and information technology specialist realized her dream in 2020, after years of searching for the keys to unlock the door. Lucille created the Texas Tribal Buffalo Project not only to fulfill the dream of being more connected with nature and her roots but to create a place where all Indigenous people could reconnect with a vital piece of their culture.[14]

An article in *Nature Conservancy Magazine* discussed the creation of the Buffalo Project: "The vision for this project started with Contreras in the years she worked with Native American activist communities in states other than Texas. Despite a long connection to the geography, Native Americans were left with only two tiny reservations in Texas."[15]

After six years helping raise bison on the Pine Ridge Indian Reservation in South Dakota, Lucille knew what she wanted: a Lipan bison ranch in Texas. She leaned into her network of Native American friends and mentors, gathering support from organizations such as the Tanka Fund and Honor the Earth, and learning from a Business of Indian Agriculture curriculum offered by the First Nations Development Institute. After praying to her ancestors for guidance and strength, Lucille applied for a USDA Beginning Farmer and Rancher loan to purchase land and begin the journey to bring the bison and Lipan Apache heritage back to Texas. She purchased seventy-seven acres of gently rolling grassland in Waelder, Texas.

AmeriCorps Vista funded eight staff positions to help grow the capacity and reach of the organization. Lucille purchased three bison from a privately owned Texas herd, an anonymous donor delivered a fourth, and The Nature Conservancy transferred five from its Zapata Ranch bison herd in Colorado.[16] (The Nature Conservancy partners with the Inter-Tribal Buffalo Council to restore bison to Native American lands from Nature Conservancy surplus bison.)

As part of the Buffalo Project, Lucille hosts summer camps to introduce Indigenous children to their heritage, which includes the bison. The *Iyanee'* Reconnections camp provides youth ages eight to fifteen with an opportunity to learn about the bison, healthy food systems, traditional plants, and traditional survival skills. They also learn about Indigenous culture through oral traditions told by Lipan Apache elders.[17]

The woman-led organization believes it is important to reconnect children with the sacred *Iyanee'* while on Native-owned land and in the care of descendants of the Buffalo Nations. "Generations of trauma—genocide, forced relocations, destruction of culture, mockery in old movies, poverty, and more—have left the Lipan Apache and other Native peoples dissociated and disconnected from the land," Contreras notes. "Our culture once revolved around the buffalo, but we'd slowly lost that connection along with the buffalo."[18]

Lucille and the Buffalo Project also provide meat to Native American people. She noted in an interview with Elene Bruess of the *San Antonio Express-News* that she feels that caring for the bison is like caring for her own relatives. That care also extends to other Apache members via harvesting a bison. Before each harvest, a ceremony takes place that connects recipients with the life and journey of the animal. "We're not ranchers in your Western European sense," Contreras said. "We're caretakers of our relatives, and we take care of them as they'll take care of us."[19]

The small herd and South Texas ranch are the starting point. Lucille represents a powerful shift in conservation to becoming more inclusive in thought and practice. She highlights the value that Native Americans have in determining the future of conservation. By creating opportunities for adults and youth to immerse themselves in and reflect on the relationship between the grasses, bison, birds, and people, she is infusing their hearts and minds with the deeper purpose of caring for the earth.

FEARLESSLY TALKING ABOUT THE ELEPHANT IN THE ROOM—CLIMATE CHANGE

Katharine Hayhoe

If you expect to keep up with Katharine Hayhoe, you had better put on your running shoes. When she is not teaching at Texas Tech University, she is trotting around the globe, talking about pollution-driven climate change. Don't worry, she uses Climate Stewards to offset her travel-related carbon emissions.[20]

Katharine grew up in an evangelical family and holds fast to her beliefs, a unique position for one of the world's leading climate experts. Entering college, she had only one path as she pursued her bachelor of science degree in physics and astronomy from the University of Toronto and a master of science and PhD in atmospheric science from the University of Illinois at Urbana-Champaign. She also was awarded honorary

Katharine Hayhoe, 2014. Photo by Ashley Rodgers. Courtesy Katharine Hayhoe and Texas Tech University–Lubbock

doctorates from Colgate University, Trinity College, and Victoria College at the University of Toronto.

Currently, she is a Paul Whitfield Horn Distinguished Professor and Endowed Chair in Public Policy and Public Law in the public administration program of the Department of Political Science at Texas Tech University. She also serves as the chief scientist for The Nature Conservancy.[21]

Katharine is a gifted writer and speaker who can weave the complexities of science and data with human emotions and needs, all with the power of an evangelical minister. While the topic of pollution-driven, human-made climate change can be depressing, Katharine always relays a positive message of hope and empowerment through individual actions. She has published 125 peer-reviewed articles, coauthored *Downscaling Techniques for High-Resolution Climate Projections*, and contributed to multiple books. She also cowrote a book with her husband, Andrew Farley, titled *A Climate for Change: Global Warming Facts for Faith-Based Decisions*. In 2021 she published her first single-author book, *Saving Us: A Climate Scientist's Case for Hope and Healing*.

Katharine doesn't just talk about the environmental impact of climate change, she considers how it affects every species, including humans. There is a deep disparity when it comes to which demographics are being the hardest hit by rising temperatures, more extreme weather events, and loss of species. She often makes the case for the economic effects that climate change has and how impoverished communities, which often contribute the least pollution, suffer the greatest, both physically and economically.

Katharine joined Oxfam's Sisters on the Planet, a "diverse group of American women who are leaders in everything from business to politics to the arts. Sisters on the Planet Ambassadors are united by a common goal: to use their influence to fight global poverty, hunger, and injustice, with a particular focus on empowering women and girls worldwide."[22] Katharine has served on various other boards and advisory teams, including the Smithsonian Natural History Museum, Earth Science Women's Network, and American Association for the Advancement of Science.[23]

While her reach might seem to expand far beyond the borders of Texas, and even the United States, she considers Texas to be one of the states that is both causing changes in the climate and suffering in real time from those effects. Increased and stronger hurricanes, tornados, and flooding events, all punctuated by extended periods of drought and extreme heat, are just some of the effects felt throughout the state, especially in disadvantaged communities.

Texas is also a garrison of climate change deniers. When Katharine made a presentation to the Austin City Council in 2016, Council Member Don Zimmerman stated, "You don't have to be as smart as a fifth grader to know that what causes the climate is the sun. I have people tell me, 'Carbon dioxide warms the earth.' No, it doesn't. The sun warms the earth, and there is more energy in our sun than humanity can comprehend." Zimmerman then insisted that the sun didn't need "a permit from the EPA" to emit solar flares. The exchange was captured by Sonia Smith of *Texas Monthly* magazine.[24]

Zimmerman's comments are similar to those heard in council chambers, commissioners' courts, state houses, and the nation's capital. As Katharine points out, when people understand the factors that create changes in the climate, that comes with a realization that we as individuals have to make changes in our daily lives and in how we conduct business. Change is hard and not something that is high on the list of the fossil fuel industry. Katharine's upbringing in a conservative, religious family and the fact that she resides in Lubbock have prepared her well to speak openly and honestly to people with a range of views.

If Katharine focused her research and public speaking only on Texas and was able to influence a reduction in carbon dioxide emissions from this state, it would affect the world. State Representative Gina Hinojosa pointed out in an article she wrote for the *Austin American-Stateman*, "If Texas were a country, we would rank seventh in the world for our carbon dioxide emissions. Major sources of these greenhouse gases are the vehicles we drive and the oil and gas industry."[25] In the article, Representative Hinojosa proceeded to lay out a climate plan that she supported for the state in 2021; the measure did not pass.

Katharine uses every medium possible to share her message and research. She created a PBS children's series called *Global Weirding*. She gives TED talks (one has more than four million views). She tweets. She produces YouTube videos. She appears on national television news shows. She speaks to religious orders. She writes research abstracts. She teaches courses at Texas Tech University. She has coauthored three national climate assessments. All the while, she remains busy as a mother, wife, daughter, and woman of faith.

Because Katharine is so prolific, I will keep this story short and urge you to read one of her books or listen to one of her talks. Her work elevates the need for conservation of lands, water, and species in Texas. Farmlands, parks, private timberlands, ranches, and wildlife management areas all sequester carbon, which helps reduce its buildup in the atmosphere and the degradation of the protective ozone layer. In addition, they mitigate the impacts of flooding and absorb the energy of hurricanes as they arrive on shore.

Katharine's legacy continues to unfold as she champions national and international issues regarding climate, the environment, and leadership roles for women in the scientific community. While Katharine's data analysis, speaking, and publications continue to influence national and international policy, she is equally content with an individual making a change in their life to reduce their carbon footprint and alter their lifestyle for the greater good of the planet. She frequently states that we have all the technology, science, and data we need—we just have to summon the will and political courage to make changes.[26]

When my mother and I were conducting interviews with the women in this book and I asked what they thought was the biggest conservation issue facing Texas. Climate change was number one, with land fragmentation, water, and population also being noted. Most saw those issues as interconnected.

ACKNOWLEDGMENTS

There is a long list of people to thank for helping make this book possible. I want to thank the women featured in the book, especially those who sent an array of photos to bring their stories to life. Thank you to Andrew Sansom and the Jacob and Terese Hershey Foundation for awarding us a grant and for all your support. I want to acknowledge David Todd for his willingness to brainstorm when I was stuck and for his generous support to manage the grant.

I am deeply grateful for the archivists who assisted us with navigating their systems to locate documents, letters, and photos to lend depth and richness to each story. I'm especially thankful for Penny Clark at Lamar University for sharing information about Bessie Reid. I now understand how important it is to have all those personal letters, documents, and photographs archived at universities and libraries across the country. I am also appreciative of the family members, including Nancy Powell Moore and Katherine Korbell Brown, who provided letters, personal stories, and photographs. Speaking of mothers, I could not have created this book if it were not for my mother. She used those dormant legal skills to conduct top-notch research, and we had a blast. From 2019 to 2022 we met several times a week at my dining room table to review the research and discussed the project.

A special giant dose of gratitude and love to my husband: You are my rock. Researching, writing, and creating a book takes hundreds of hours of working alone in silence—a hard feat to achieve while we were both working at home during the pandemic. But you made it happen and cheered me on.

A big thanks for all the wisdom and encouragement that Ellen Temple provided during the process. An even bigger thanks to her daughter, and the best friend a woman could ever have, Susie Temple. The isolation of your river cabin helped get me over the hump. I also want to say thanks to fellow author and friend, James Fenelon. I cherish our irreverent chats.

Thank you to Shannon Davies for encouraging me to tell this story, and to everyone at Texas A&M University Press and Texas Woman's University for taking on the project. A special thanks to Pam Nelson Harte for all your advice, kindness, and for supporting this project. I am honored to have the support of the Kathie and Ed Cox Jr. Books on Conservation Leadership and the Nancy and Ted Paup series on Pioneering Women: Leaders and Trailblazers.

NOTES

Chapter 1. Saving the Birds

1. "Who Are We," General Federation of Women's Clubs, accessed May 8, 2025, https://www.gfwc.org/who-we-are/.
2. "Save the Birds," *Galveston Daily News*, December 18, 1893, 3.
3. Paul R. Ehrlich et al., "Plume Trade," Stanford Birds, 1988, https://web.stanford.edu/group/stanfordbirds/text/essays/Plume_Trade.html.
4. "Meeting Directory," in *Morrison and Fourmy's General Directory of the City of Galveston* (Morrison and Fourmy, n.d.), 297.
5. Dayton Duncan and Ken Burns, *Blood Memory: The Tragic Decline and Improbable Resurrection of the American Buffalo* (Alfred A. Knopf, 2023), 196.
6. Stanley D. Casto, "Origin of the Audubon Society in Texas," *Bulletin of the Texas Ornithological Society* 18 (1985): 2–5.
7. Becky Little, "How the Galveston Hurricane of 1900 Became the Deadliest U.S. Natural Disaster," History.com, August 29, 2017, https://www.history.com/news/how-the-galveston-hurricane-of-1900-became-the-deadliest-u-s-natural-disaster.
8. Casto, "Origin of the Audubon Society."
9. Jesse Greenspan, "The History of Audubon," Audubon, June 4, 2018, https://www.audubon.org/about/history-audubon-and-waterbird-conservation.
10. H.C.R. 88, 76th Texas Legislature (1999).
11. "Mrs. Bruce Reid, Naturalist, Dies," *Fort Worth Star-Telegram*, April 29, 1962.
12. Decca Lamar West, "State Federation Interested in Federal Migratory Bird Law," *Austin American-Statesman*, November 29, 1914, 15.
13. Mrs. Bruce Reid [Bessie Reid], "Bird Protection Urged," *Austin American-Statesman*, May 3, 1914, 17.
14. Mrs. Bruce Reid [Bessie Reid], "State Federation Interested in Federal Migratory Bird Law," *Austin American-Statesman*, November 29, 1914, 16.
15. "Mrs. Bruce Reid to Speak in Houston," *Houston Post*, October 19, 1917, 10.
16. Valvera Hampton, "Woman Naturalist's Home Haven for Texas Birds," *Houston Chronicle*, May 24, 1931, 4–5.

17. Bessie Reid to Robert T. Allen, July 17, 1934, Audubon Florida Records, 1900–1970, Box 2, Folder 6: Texas Letters and Misc., 1934–1938, Digital Commons @ University of South Florida, https://digitalcommons.usf.edu/audubon_fl_records/38/.
18. "Crash Kills Two Birds: Naturalist Has Wreck En Route to Waco; Snakes Saved," *Waco Tribune-Herald*, December 17, 1933, 2.
19. Lisa Gonzalez, interview by Jennifer Bristol, September 12, 2023.
20. Karen Harden McCracken, *Connie Hagar: The Life History of a Texas Birdwatcher* (Texas A&M University Press, 1986), 43.
21. Pete A. Y. Gunter, "R.E. Jackson and the Early Big Thicket Conservation Movement, 1929–1957," *East Texas Historical Journal* 37, no. 2 (1999): 53–63, https://scholarworks.sfasu.edu/cgi/viewcontent.cgi?article=2274&context=ethj.
22. Elizabeth Kendall, "Beautification of Highways Being Urged," *Austin American-Statesman*, November 26, 1933.
23. "A Report of the Special Meeting of the Texas Academy of Science, Beaumont, June 25–27, 1937," July 15, 1937, Audubon Florida Records, 1900–1970, Box 2, Folder 6: Texas Letters and Misc., 1934–1938, Digital Commons @ University of South Florida, https://digitalcommons.usf.edu/audubon_fl_records/38/.
24. Eldon Stephen Branda, "Dobie, Bertha McKee (1890–1974)," Texas State Historical Association, 1976, last updated December 1, 1994, https://www.tshaonline.org/handbook/entries/dobie-bertha-mckee.
25. Center for History and Culture of Southeast Texas and the Upper Gulf Coast, "Bessie Reid was an inveterate collector. While traveling across Southeast Texas in search of unusual flora and fauna, she also gathered local Native American folklore. One story of star-crossed Atakapan lovers who drowned in Sabine Lake inspired Reid to pen 'The Legend of Kisselpoo,'" Facebook, March 27, 2020, https://www.facebook.com/centerhistoryculture/posts/1282886281917295?_rdr.
26. Milton Turner, "'Nut Hill' on Village Creek Near U.S. 96 Is Natural Habitat of Plants, Birds; Mrs. Bessie Reid, Husband Foster Unspoiled Paradise to Study Texas Folklore," *Sunday Enterprise* (Beaumont, TX), May 19, 1957.
27. "Audubons Invite Guest Speakers," *Austin American-Statesman*, April 13, 1958, 38.
28. "Conservation Groups Announce Land Purchase to Preserve," *El Paso Times*, January 19, 1975.
29. James Herzberg, "Naturalist Armand Yramategui, 1923–1970," *Houston Review* 17, no. 1 (1995): 49–58, https://houstonhistorymagazine.org/wp-content/uploads/2014/02/17.1-Naturalist-Armand-Yramategui-1923-1970-James-Herzberg.pdf.
30. "Big Thicket Awarded," *Liberty Vindicator* (Nacogdoches, TX), October 10, 1974, 17.
31. "Married at Kountze," *Galveston Daily News*, June 24, 1892, 5.
32. "Venerable Mrs. Hooks, Sr., Widely Known Naturalist, Succumbs in Texas at 86," *Weekly News* (Marksville, LA), January 17, 1963.

33. Family letters and memories provided by Bonnie Korbell, Katherine Brown, Joanne Pfeil Boyce, and William Edward Hooks Jr., 2021.
34. "Notes on Those Attending A. & M. Short Course," *Bryan (TX) Daily Eagle*, July 28, 1930, 3.
35. Family letters and memories.
36. Judith Walker Linsley and Ellen Walker Rienstra, "McLean, Marrs (1883–1953)," Texas State Historical Association, May 1, 1995, last updated July 20, 2016, https://www.tshaonline.org/handbook/entries/mclean-marrs.
37. Melanie Wiggins, *They Made Their Own Law: Stories of Bolivar Peninsula* (Rice University Press, 1990), 99.
38. Family letters and memories.
39. Family letters and memories.
40. "'Experts' Proved Wrong on Texas Bird Life Book," *Lubbock Avalanche-Journal*, September 22, 1974.
41. Family letters and memories.
42. McCracken, *Connie Hagar*.
43. McCracken, *Connie Hagar*, 149.
44. McCracken, *Connie Hagar*, 89.
45. Natalie Moore, "The Bird Lady of Rockport," Texas Highways, November 1, 2009, https://texashighways.com/travel-news/the-bird-lady-of-rockport/.
46. "No Mere Bird Watcher: Rockport Ornithologist Gets Overdue Recognition," *Corpus Christi Times*, September 2, 1956, 7.
47. Penny Peckenpaugh, "Mrs. Jack Hagar Stresses Need for Conservation," *Corpus Christi Caller-Times*, October 24, 1948.
48. "Girl Scout Heads to Hear Talk on Sea Flora, Fauna," *Corpus Christi Times*, February 20, 1951, 15.
49. "Special Recognition Day for Mrs. Jack Hagar at Corpus Christi College," *Corsicana (TX) Daily Sun*, February 25, 1960.
50. "Texas Coast Getting Audubon Society's Conference," *Fort Worth Star-Telegram*, November 4, 1962, 6.
51. "Former Corsicanan Was Noted for Her Work in Ornithology," *Corsicana (TX) Daily Sun*, December 7, 1973.
52. Madge Lindsay, interview by Jennifer Bristol, February 19, 2020.
53. Lindsay, interview.
54. *Nature Tourism in the Lone Star State: Economic Opportunities in Nature* (State Task Force on Texas Nature Tourism, 1994).
55. *Making Nature Your Business*, Texas Parks and Wildlife Department, 2002, 7.
56. *Making Nature Your Business*, 3–9.
57. Dan Parker, "Birding Trail's First Site Opens Friday in Rockport," *Corpus Christi Caller-Times*, September 7, 1995.
58. Lindsay, interview.

59. Shelley Plante, interview by Jennifer Bristol, 2021.
60. *The Great Texas Birding Classic: Celebrating a Decade of Avian Habitat Conservation* (Gulf Coast Bird Observation and the Texas Parks and Wildlife Department, 2005).
61. Gillian Swanson, "Mid-Valley City Shows Interest in Birding Center," *Monitor* (McAllen, TX), September 25, 1997.
62. Cheryl Smith, "City Antes Up Land for Birding HQ: Offer Hammered Out Hours Prior to Deadline for Lucrative Center," *Monitor* (McAllen, TX), December 23, 1998, 15.
63. Cheryl Smith, "World Birding Center Will Nest in Mission," *Monitor* (McAllen, TX), April 30, 1999.
64. "Bentsen–Rio Grande Valley State Park," Texas Parks and Wildlife Department, accessed May 8, 2024, https://tpwd.texas.gov/state-parks/bentsen-rio-grande-valley.
65. Allen Essex, "Weslaco Proceeding with $1 Million Bond Issue for World Birding Center," *Monitor* (McAllen, TX), November 24, 2000.
66. Carter Smith, interview by Jennifer Bristol, December 8, 2021.
67. "Estero Llano Grande State Park," Texas Parks and Wildlife Department, accessed May 8, 2024, https://tpwd.texas.gov/state-parks/estero-llano-grande/history.
68. Martha Garcia, interview by Jennifer Bristol, November 1, 2022.
69. "Madge Lindsay Biographical Profile and Resume." 2003.
70. Clifford Eugene Shackelford et al., *Hummingbirds of Texas: With Their New Mexico and Arizona Ranges* (Texas A&M University Press, 2005).
71. Lindsay, interview.
72. Lindsay, interview.
73. Gina Donovan, "Houston Audubon Celebrates Forty Years in Conservation!," *Naturalist*, September/October 2009, 1–2.
74. Winnie Burkett, interview by David Todd, October 14, 1999, Texas Legacy Project.
75. Joy Hester, interview by Valarie Bristol and Jennifer Bristol, July 22, 2019.
76. Hester, interview.
77. Hester, interview.
78. Hester, interview.
79. Hester, interview.
80. Helen Drummond, interview by Jennifer Bristol, December 22, 2022.
81. Drummond, interview.
82. Drummond, interview.
83. US Census, Houston, January 2023, https://www.census.gov/quickfacts/houstoncitytexas.
84. Beth Boone, press release, Houston Audubon, April 8, 2019.
85. Mayor's Office, City of Houston, "The City of Houston Proclaims 'Lights Out Nights' in Houston to Help Migrating Birds Traveling Through Region," press release, April 12, 2021.

86. Drummond, interview.
87. Drummond, interview.
88. Drummond, interview.

Chapter 2. Creating Parks and Preserves

Portions of this chapter appeared in George Bristol, *Texas State Parks the First 100 Years: 1923-2023* (TCU Press, 2023).

1. Mrs. Grady Rash Jr., "Daughters of the Republic of Texas," Texas State Historical Association, 1952, last updated February 26, 2025, https://www.tshaonline.org/handbook/entries/daughters-of-the-republic-of-texas.
2. Elizabeth Hayes Turner, "Ballinger, Betty Eve (1854–1936)," Texas State Historical Association, 1976, last updated July 26, 2021, https://www.tshaonline.org/handbook/entries/ballinger-betty-eve.
3. "About Us- DRT History," Daughters of the Republic of Texas, accessed May 13, 2025, https://www.drtinfo.org/Members/Members/About-Us/DRT-History.aspx?hkey=751fb335-3b40-4443-a640-d706b5dd2b1f.
4. Bristol, *Texas State Parks the First 100 Years*, 21.
5. Megan Seaholm, "Texas Federation of Women's Clubs," Texas State Historical Association, 1976, last updated April 4, 2022, https://www.tshaonline.org/handbook/entries/texas-federation-of-womens-clubs.
6. "TAWC," Woman's Collection, Texas Woman's University Libraries, accessed May 13, 2025, https://twu.edu/library/womans-collection/collections/civic-organizations/tawc/.
7. "Timeline of Voting and Elections in Texas," League of Women Voters of Texas, accessed March 5, 2022, https://lwvtexas.org/content.aspx?page_id=22&club_id=979482&module_id=485202.
8. "Palestine to Houston Highway Campaign by Post Party," *Houston Post*, June 20, 1915, 44.
9. Martha Deeringer, *Flower of the Wilderness: Mother Neff and the State Parks of Texas* (BookLocker.com, 2014).
10. Thomas E. Turner, "Neff, Pat Morris (1871–1952)," Texas State Historical Association, 1976, last updated February 17, 2022, https://www.tshaonline.org/handbook/entries/neff-pat-morris.
11. James Wright Steely, *Parks for Texas: Enduring Landscapes of the New Deal* (University of Texas Press, 1999).
12. "Women Urged to Vote for Amendment," *Courier-Gazette* (McKinney, TX), July 2, 1923.
13. Steely, *Parks for Texas*, 6

14. "State Park Board on Another Trip to Examine New Sites," *Eagle* (Bryan, TX), August 21, 1924, 1.
15. Allison Baughman, "Marian 'Mamie' Rather Powell," East Texas History, accessed July 2019, https://easttexashistory.org/items/show/22.
16. Rachel M. Gunter, "Neal, Margie Elizabeth (1875–1971)," Texas State Historical Association, 1976, last updated August 8, 2022, https://www.tshaonline.org/handbook/entries/neal-margie-elizabeth.
17. D. E. Colp to Mrs. E. A Bellis, February 12, 1927. Courtesy Nancy Powell Moore
18. Mrs. Ben Powell to D. E. Colp, February 19, 1927. Courtesy Nancy Powell Moore.
19. D. E. Colp to Mrs. Ben Powell, February 21, 1927. Courtesy Nancy Powell Moore.
20. Mrs. Ben Powell to Mrs. R. D. Rather, March 2, 1927. Courtesy Nancy Powell Moore.
21. D. E. Colp to Mrs. E. A. Bellis, March 17, 1927. Courtesy Nancy Powell Moore.
22. "State Accepts 24 New Parks, Mrs. Ben Powell, Austin, Presses Resolution in Legislature," *Austin American-Statesman*, March 19, 1927, 2.
23. D. E. Colp and the State Park Association to Miss Julia D. Owens, May 7, 1927. Courtesy Nancy Powell Moore.
24. Nancy Powell Moore, interview by Jennifer Bristol, February 27, 2021.
25. June M. Steele, "Phebe Warner: Community Building in the Texas Panhandle, 1898–1935" (MA thesis, Texas Tech University, 2000).
26. Hugh Allen Anderson, "Warner, Phebe Kerrick (1866–1935)," Texas State Historical Association, 1952, last updated August 1, 1995, https://www.tshaonline.org/handbook/entries/warner-phebe-kerrick.
27. Phebe Warner, weekly column, *San Angelo Daily Standard—Western Weekly*, June 1, 1924, 24.
28. Phebe Warner, weekly column, *San Angelo Daily Standard—Western Weekly*, June 28, 1925.
29. Phebe Warner, weekly column, *Southwest Plainsman and Panhandle Weekly* (Amarillo, TX), November 21, 1930.
30. "Mrs. Phebe K. Warner," *Fort Worth Star-Telegram*, May 18, 1935, 4.
31. "Transformation of Cave Springs Sabine Bottoms, Narrative of What One Woman Can Do Who Has a Well-Defined Plan and Follows It," *Galveston Daily News*, April 24, 1921, 15.
32. "Services Monday for Mrs. Martin," *Hood County News-Tablet* (Granbury, TX), January 9, 1958, 1.
33. Mrs. W. C. Martin [Florence George Martin], "State Park System Is Needed to Further Move of Acquainting Texans with State," *Fort Worth Star-Telegram*, July 13, 1924, 29.
34. "Caddo Lake Park to be Suggested," *Fort Worth Star-Telegram*, August 19, 1928, 3.
35. Neel Baumgardner, "A State Park System for Texas: 100 Years of Public Conservation," Texas Land Conservancy, accessed May 13, 2025, https://www.texaslandconservancy.org/tlc-stories/a-state-park-system-for-texas.

36. "Mrs. J. F. Welder Succumbs at 94," *Victoria (TX) Advocate*, September 10, 1958, 1.
37. *Survey of Ecosystems Which Should Be Considered Sensitive in Planning the Future Development of the City of Austin* (Travis Audubon, 1972).
38. Janet Poage, personal document, "History of Now or Never," January 20, 1986. St. Edwards University, Wild Basin Creative Research Center archives.
39. Mark Bierner, "Janet Poage: Reflections on Now or Never and the Wild Basin," Wild Basin Preserve, n.d. St. Edwards University, Wild Basin Creative Research Center archives.
40. Austin Creeks, Horizons '76 printed publication for the City of Austin, 1976.
41. Martha Hudson and Flo Macklin, interview by Jennifer Bristol and Valarie Bristol, September 11, 2019.
42. Hudson and Macklin, interview.
43. Roberta S. Duncan, "McCallum, Jane Legette Yelvington (1877–1957)," Texas State Historical Association, 1976, last updated July 30, 2020, https://www.tshaonline.org/handbook/entries/mccallum-jane-legette-yelvington.
44. "Jane Y's Petticoat Lobby," *Austin American-Statesman*, March 2, 1986, 49.
45. Janet Poage, "Snowball in Hell, or How Seven Little Old Ladies in Tennis Shoes Whipped the Establishment and Created an Urban Nature Preserve in Austin, Texas," unpublished manuscript, 1992.
46. Poage, "Snowball in Hell."
47. Hudson and Macklin, interview.
48. Suzanne Majors Davis, "Rollingwood's Flo Macklin, 93, Was Nationally Honored for Her Volunteerism," *Westlake (TX) Picayune*, September 2021.
49. Hudson and Macklin, interview.
50. "Lucille A. Brightbill Stegman," *Austin American-Statesman*, November 3, 2007, 24.
51. Poage, "Snowball in Hell."
52. Linda Kerr, "Fund Drive Succeeds, Wild Basin Backers Raise Needed $175,000," *Austin American-Statesman*, July 1, 1977.
53. Greg Olds, "Wild Basin Awarded Federal Funds," *Westlake (TX) Picayune*, October 20, 1978.
54. Poage, "Snowball in Hell."
55. Poage, "Snowball in Hell."
56. "About Wild Basin," St. Edward's University, accessed 2022, https://www.stedwards.edu/wild-basin/about.
57. Jeanne Claire Van Ryzin, "Physician Fascinated with Culture and Nature," *Austin American-Statesman*, January 11, 2005, 15.
58. Michael Barnes, "Jane Sibley, 95, Saved the Austin Symphony," *Austin American-Statesman*, May 22, 2019.
59. Jane Dunn Sibley, *Jane's Window: My Spirited Life in West Texas and Austin*, as told to Jim Comer (Texas A&M University Press, 2013).
60. Sibley, *Jane's Window*.

61. Sibley, *Jane's Window.*
62. "Seminole Canyon State Park and Historic Site," Texas Parks and Wildlife Department, April 7, 2022, https://tpwd.texas.gov/state-parks/seminole-canyon.
63. Jeff Turpin, "Texas Art Patrons Gather to Save Rock Site," *Del Rio (TX) News-Herald*, May 12, 1991, 11.
64. "About," Sibley Nature Center, accessed 2020, https://sibleynaturecenter.org/about.

Chapter 3. Planting the Seeds

1. Magnolia Neighbors Greene, *History of Texas Garden Clubs, Inc., 1928–1948* (Texas Garden Clubs, 1949). 7.
2. "Buds Planted in County Park," *Fort Worth Star-Telegram*, February 5, 1933, 5.
3. "Mrs. Henry B. Trigg Named Member Board of Big Bend Park," *Corsicana Daily Sun*, August 29, 1940, 4.
4. John Jameson, *The Story of Big Bend National Park* (University of Texas Press, 1996).
5. "Mrs. Louis J. Wardlaw on Board for Big Bend," *Fort Worth Star-Telegram*, September 20, 1940, 11.
6. "Talk, Pictures on Big Bend on Rotary Club's Program," *Fort Worth Star-Telegram*, October 5, 1941, 7.
7. "Big Bend Park Given Boost, Mrs. Wardlaw Heard at Harlingen," *Brownsville (TX) Herald*, March 13, 1941, 15.
8. Nella Mae Dieter, "Garden Clubbers Hope to Save Bluebells from Buffalo Fate," *Austin American-Statesman*, July 22, 1945, 24.
9. Pat Ford, "President Tells of Widened Field," *Valley Morning Star* (Harlingen, TX), March 4, 1948, 1.
10. Annie Lee Shelton, "Back to Nature," *Houston Chronicle*, July 10, 1949, 110.
11. "Audubon Society Slates First Meeting of Year," *Austin American-Statesman*, September 16, 1953, 2.
12. "Ex-State Chief of Garden Clubs Dies at Austin," *Fort Worth Star-Telegram*, September 13, 1965, 2.
13. Mrs. Howard S. Kittel, *Stepping Stones for Garden Clubs: 1963–1965* (Courtesy of the Sears-Roebuck Foundation, 1965), 102.
14. *Water Pollution Control and Abatement: Hearings Before a Subcommittee of the Committee on Government Operations, Part 6—Texas*, 88th Cong. 3310 (1963).
15. "Resident Gives Officials Book," *Fort Worth Star-Telegram*, October 17, 1964, 11.
16. Howard Bryan, "Garden Clubs Seek Image Change," *Albuquerque Tribune*, April 22, 1974, 1.
17. "Flower Day Set," *Fort Worth Star-Telegram*, March 2, 1984, 35.
18. "Sanitation Drive," *Fort Worth Star-Telegram*, July 6, 1949, 5.
19. "Negro Observers May Attend Garden Meetings," *Corpus Christi Caller-Times*, September 22, 1951, 20.

20. "Negro Women Organize Club to Further Garden Interests," *Austin American-Statesman*, April 11, 1954, 67.
21. Lewis L. Gould, *Lady Bird Johnson: Our Environmental First Lady* (University Press of Kansas, 1988), 55.
22. Julia Sweig, *Lady Bird Johnson: Hiding in Plain Sight* (Random House, 2021), 138.
23. *Lone Star Gardener*, Fall 2022.
24. Bess Carroll Woolford and Ellen Schulz Quillin, *The Story of the Witte Memorial Museum, 1922–1960* (San Antonio Museum Association, 1966), 5–6.
25. Woolford and Quillin, *Story of the Witte*, 12.
26. Ellen D. Schulz, *Texas Wild Flowers: A Popular Account of the Common Wild Flowers of Texas* (Laidlaw Bros., 1928).
27. Ellen Schulz Quillin and Robert Runyon, *Texas Cacti: A Popular and Scientific Account of the Cacti Native of Texas* (Texas Academy of Science, 1930).
28. Woolford and Quillin, *Story of the Witte*, 12.
29. "Museum Fund Left by San Antonio Man," *Fort Worth Star-Telegram*, October 16, 1925, 19.
30. "Museum Expedition to Explore Big Bend Area," *Fort Worth Star-Telegram*, August 17, 1929, 13.
31. "Ellen Quillin Day Slated Here Sunday," *San Antonio Express-News*, October 30, 1960, 12.
32. Cecilia Steinfeldt, "Quillin, Ellen Dorothy Schulz (1887–1970)," Texas State Historical Association, June 1, 1995, last updated February 22, 2018, https://www.tshaonline.org/handbook/entries/quillin-ellen-dorothy-schulz.
33. "Quillin Egg Collection," Welder Wildlife Foundation, accessed May 14, 2025, https://welderwildlife.org/quillin-egg-collection/.
34. "About Mrs. Johnson," Lady Bird Johnson Wildflower Center, accessed 2020, http://www.ladybirdjohnson.org/about.
35. Lewis L. Gould, *Lady Bird Johnson: Our Environmental First Lady* (University Press of Kansas, 1988), 96.
36. Gould, *Lady Bird Johnson*, 100.
37. Gould, *Lady Bird Johnson*, 122.
38. Rob McCorkle, "Park Reflects Personality of President," *Odessa (TX) American*, January 16, 2000.
39. Claudia Hazelwood, "Packsaddle Mountain Fight," Texas State Historical Association, 1952, last updated January 7, 2021, https://www.tshaonline.org/handbook/entries/packsaddle-mountain-fight.
40. Danna McDonald, "The Sale of Enchanted Rock," research document (Texas Parks and Wildlife Archeology Library, 1978).
41. McDonald, "The Sale of Enchanted Rock."
42. Gould, *Lady Bird Johnson*, 125.
43. "Native Plants of North America," Lady Bird Johnson Wildflower Center, July 1, 2022, https://www.wildflower.org/plants-main.

44. Ellen Temple, interview by Jennifer Bristol and Valarie Bristol, 2019.
45. Erica Whittington, "A Conversation with Ellen Temple," Humanities Texas, October 2013, https://www.humanitiestexas.org/news/articles/conversation-ellen-temple.
46. "Citizens at Last," Women in Texas History, accessed June 16, 2025, https://www.womenintexashistory.org/video/citizens-at-last/.
47. Ellen Temple, interview.
48. Ellen Temple, interview.
49. Ellen Temple, interview.
50. Ellen Temple, interview.
51. "The Stephen F. Austin Pineywoods Native Plant Center: 1998–2012," Stephen F. Austin State University, accessed August 22, 2022, https://www.sfasu.edu/docs/sfa-gardens/history-native-plant-center.pdf.
52. "Donor Spotlight: Ellen Temple," *Caesar Kleberg Tracks*, Spring 2022.
53. Susan Temple, interview by Jennifer Bristol, 2022.
54. *Philosophical Society of Texas: Proceedings of the Annual Meeting at Austin, November 30–December 2, 2001* (Philosophical Society of Texas, 2002).
55. Ellen Temple, interview.
56. Jonathan K. Gerland, *Boggy Slough: A Forest, a Family, and a Foundation for Land Conservation* (Texas A&M University Press, 2022).
57. Jesse Woche and Julie Shackelford, "History of the Neches River National Wildlife Refuge," Conservation Fund archives, October 28, 2019.
58. "Ellen Temple Named East Texan of the Year," *KTRE News*, May 28, 2009.
59. "Texas Native Seeds Program Expands Further in Pursuit of Statewide Impact," Caesar Kleberg Wildlife Research Institute, Texas A&M University–Kingsville, August 8, 2018, https://www.ckwri.tamuk.edu/publications/texas-native-seeds-program-expands-further-pursuit-statewide-impact.
60. "Donor Spotlight: Ellen Temple."

Chapter 4. Saving East Texas Forests and Rivers

1. Maxine Johnston, interview by Jennifer Bristol and Valarie Bristol, July 2019.
2. Johnston, interview.
3. Peter A. Y. Gunter, "R. E. Jackson and the Early Big Thicket Conservation Movement: 1929–1957," *East Texas Historical Journal* 37, no. 2 (1999): 53–63.
4. Ellen Temple, video interview, Texas Women in Conservation Terry Hershey Award, Audubon Texas, 2018.
5. Johnston, interview.
6. James J. Cozine Jr., *Saving the Big Thicket: From Exploration to Preservation, 1685–2003* (Big Thicket Association / University of North Texas Press, 2004), 142.
7. Johnston, interview.

8. Bill Choyke, "Woods Still Thicket of Problems," *Fort Worth Star-Telegram*, July 24, 1978, 11.
9. Lorraine G. Bonney, Maxine Johnston, and Peter A. Y. Gunter, *The Big Thicket Guidebook: Exploring the Backroads and History of Southeast Texas* (Big Thicket Association / University of North Texas Press, 2011), 18–19.
10. Bonney, Johnston, and Gunter, *Big Thicket Guidebook*.
11. Johnston, interview.
12. Johnston, interview.
13. Peter A. Y. Gunter, "The Big Thicket as History and Ecology" in Full / Official Symposium Proceedings, Native Plant Society of Texas, October 16–19, 2008, 19.
14. Ellen Buchanan, interview by Jennifer Bristol, January 2023.
15. Buchanan, interview.
16. Janice Bezanson, interview by Jennifer Bristol and Valarie Bristol, 2019.
17. Janice Bezanson, interview.
18. Janice Bezanson, interview.
19. Janice Bezanson, interview.
20. Janice Bezanson, interview.
21. Wendee Holtcamp, "Texas Thirst for Dams Bucks National Trend," *Pacific Standard*, June 17, 2017.
22. Janice Bezanson, interview by David Todd, Texas Legacy Project, January 20, 2007, https://www.texaslegacy.org/narrator/janice-bezanson/.
23. Committee for Water for East Texas's Future, "Who Is Janice Bezanson?," *Longview News-Journal*, n.d.
24. Janice Bezanson and Stephen Bezanson, "I Am Janice Bezanson," *Longview News-Journal*, n.d.
25. Janice Bezanson, interview by J. Bristol and V. Bristol.
26. Jesse Woche and Julie Shackelford, "History of the Neches River National Wildlife Refuge," Conservation Fund archives, October 28, 2019.
27. Woche and Shackelford, "History of the Neches."
28. Woche and Shackelford, "History of the Neches."
29. "Julie Shackelford," Conservation Fund, accessed May 15, 2025, https://www.conservationfund.org/our-experts/julie-shackelford.
30. Janice Bezanson, interview by J. Bristol and V. Bristol.
31. "About TCA," Texas Conservation Alliance, accessed May 15, 2025, https://www.tcatexas.org/about-us.
32. David Bezanson, interview by Jennifer Bristol, 2021.
33. David Bezanson, interview.
34. Texas Land Conservancy, home page, accessed 2022, https://www.texaslandconservancy.org/.
35. James A. Frazier (1819–1889), census and military records, accessed June 24, 2025, https://www.ancestry.com/family-tree/person/tree/204893292/person/202680131874/facts.

36. Thad Sitton, "Freedmen's Settlements," Texas State Historical Association, July 17, 2007, last updated February 1, 2022, https://www.tshaonline.org/handbook/entries/freedmens-settlements.
37. Texas Freedom Colonies Project, home page, accessed May 15, 2025, https://www.thetexasfreedomcoloniesproject.com/.
38. Andrea Wright, "Legacy of Land," *Beaumont (TX) Enterprise*, May 22, 2022.
39. "Relative of Freed Slave Donates Zavala Land in Memory of Grandfather," *Monitor* (McAllen, TX), May 10, 2002, 20.
40. Richard M. Donovan, *Paddling the Wild Neches* (Texas A&M University Press, 2006).

Chapter 5. The Policymakers

1. *Spirit Waters* (Beverly S. Sheffield Education Center, May 27, 2020), https://www.austintexas.gov/sites/default/files/files/Parks/ANSC/2020%20Spirit%20Waters%20Booklet%20hyperlinked%20table%20of%20contents%205.27.2020.pdf.
2. "History," Pearl Brewery, accessed May 15, 2025, https://atpearl.com/about/history/.
3. Katie Hennessey, "Sisters of Charity of Incarnate Word Sign Conservation Easement to Protect San Antonio River Headwaters," *San Antonio Current*, June 5, 2022, https://www.sacurrent.com/news/sisters-of-charity-of-incarnate-word-sign-conservation-easement-to-protect-san-antonio-river-headwaters-24130659.
4. Micaela Valadez, "Sinkin, Fay M. Bloom (1918–2009)," Texas State Historical Association, February 22, 2022, https://www.tshaonline.org/handbook/entries/sinkin-fay-m-bloom.
5. Carmina Danini and Edmund Tijerina, "A Guardian of the Earth and Its People," *San Antonio Express-News*, March 6, 2009, 1B.
6. Danini and Tijerina, "Guardian of the Earth."
7. Jarboe Russell, "The Fay Way Made a Difference," *San Antonio Express-News*, March 15, 2009, 9B.
8. "Timeline," Edwards Aquifer Authority, accessed May 15, 2025, https://www.edwardsaquifer.org/eaa/history/eaa-timeline/.
9. Endangered Species Act, Pub. L. No. 93-205, 87 Stat. 884 (1973).
10. "Purchase Plan Pushed," *San Antonio Light*, September 25, 1974, 22.
11. Ben King Jr., "Most of Aquifer Unzoned," *San Antonio Express*, December 15, 1975, 4.
12. Rick Casey, "History 101: What Voters Said About Aquifer Development," *San Antonio Express-News*, January 20, 2002, 3A.
13. King, "Most of Aquifer Unzoned."
14. *San Antonio and Baxar County Report on Reclamation and Re-Use of Municipal Wastewater* (Freese, Nichols, Endress Consulting Engineers, 1971).

15. "Berriozabal Has Strong Sense of Responsibility to Native District," *San Antonio Light*, June 26, 1981, 29.
16. David Todd and David Weisman, eds., *The Texas Legacy Project: Stories of Courage and Conservation* (Texas A&M University Press, 2010), 47.
17. "Helen Dutmer Obituary—San Antonio, TX," Dignity Memorial, accessed May 15, 2025, https://www.dignitymemorial.com/obituaries/san-antonio-tx/helen-dutmer-9007425.
18. Todd and Weisman, *Texas Legacy Project*, 48.
19. Deborah Weser, "Debate over Water Persists as Reservoir Funds Grows," *San Antonio Light*, July 11, 1984, 19.
20. James Martinez, "Citizens Told to Protect Aquifer Zone," *San Antonio Light*, June 30, 1987, 16.
21. John Tackett, "Reservoir Opponents Meet with Engineers," *San Antonio Light*, February 9, 1991, 35.
22. Patrick Crimmins and Susie Phillips Gonzales, "Applewhite Hot Potato on Council Menu Tonight," *San Antonio Light*, May 9, 1991, 1.
23. Jim Wood and Susie Phillips Gonzales, "City Council Pulls Plug on Applewhite Reservoir Emergency Meeting of Water Board Today," *San Antonio Express-News*, May 10, 1991, 1A.
24. "Timeline," Edwards Aquifer Authority.
25. Kirstin Downy, "SL Unit's Remnant Up for Sale," *Washington Post*, October 21, 1989.
26. "Task Force Leader Works Hard to Preserve Water," *San Antonio Express-News*, April 2, 1991.
27. John MacCormack, "Gets Go Ahead," *San Antonio Express-News*, May 21, 1993, 1A.
28. Lydia Saldaña, "A Man for All Seasons: Hixon Led Efforts to Preserve Government Canyon," *Texas Parks and Wildlife Magazine*, November 2019.
29. Ralph Winingham, "Expansion of Natural Area OK'd," *San Antonio Express-News*, January 22, 1999, 1B.
30. Christopher Anderson, "Conner Fends Off Two Others to Replace Outgoing Marbut," *San Antonio Express-News*, May 1, 1999.
31. Jerry Needham, "City Aims to Protect Aquifer Proposition 3 Would Fund Land Purchases," *San Antonio Express-News*, April 15, 2000.
32. Valarie Bristol, interview by Jennifer Bristol, October 4, 2022.
33. Christopher Anderson, "Council Votes to Buy Sensitive Rural Land," *San Antonio Express-News*, December 15, 2000, 8B.
34. Jaime Castillo, "PGA Dishes Up Political Hot Potato, City Council Members Perform a Balancing Act," *San Antonio Express-News*, February 3, 2002.
35. "TPC: Course Rise from Ruins," *Austin American-Statesman*, September 15, 2005, 30.
36. Jerry Needham, "Extension Would Enable the City to Buy Up More Land to Protect the Edwards Aquifer," *San Antonio Express-News*, March 11, 2003.

37. Colin McDonald, "Government Canyon Adds Protected Area," *San Antonio Express-News*, May 29, 2009, 1A.
38. Garry Merritt, interview by Jennifer Bristol, May 13, 2023.
39. Michael Barnes, "Conservation Rock Star," *Austin American-Statesman*, December 8, 2013, D11.
40. Susan Rieff, interview by David Todd and David Weisman, Texas Legacy Project, February 21, 2008, https://www.texaslegacy.org/narrator/susan-rieff/.
41. Cathy Frisinger, "Susan Rieff '78 Worked to Protect Our Environment Long Before It Came in Vogue," *TCU Magazine*, 2008.
42. Peter Elkind, "The Buck Stops Here," *Texas Monthly*, January 1990, 142–43.
43. "Are Your Water Wells Safe?," *Jeff Davis County News*, April 20, 1989, 7.
44. "Texas Candidate's Comment About Rape Causes a Furor," *New York Times*, March 26, 1990, 15.
45. Rieff, interview by Todd and Weisman.
46. Frisinger, "Susan Rieff."
47. Rieff, interview by Todd and Weisman.
48. Rieff, interview by Todd and Weisman.
49. "Flowchart of TCEQ and Its Predecessor Agencies," Texas Commission on Environmental Quality, accessed October 8, 2022, https://www.tceq.texas.gov/agency/organization/tceq-flowchart.html.
50. Ralph K. M. Haurwitz, "Group Questions Delay in Salamander Listing," *Austin American-Statesman*, June 25, 1996, 5.
51. "Money to Buy Land for Balcones Refuge Is Reduced," *Austin American-Statesman*, January 19, 1999, 30.
52. Barnes, "Conservation Rock Star."
53. Rieff, interview by Todd and Weisman.
54. Bill McCann, "Save Rare Species Group Says," *Austin American-Statesman*, February 26, 1988, 20.
55. Bill McCann, "Endangered Species Ordinance Unveiled," *Austin American-Statesman*, September 2, 1988, 20.
56. Janet Wilson, "Travis Seeks Modification of Road Pact: County Officials Fear Suit in Vireo-Stalled Widening," *Austin American-Statesman*, November 26, 1988, 29.
57. Laylan Copelin, "Environmentalist Slam Bills, Legislation Would Limit Effort to Protect Species," *Austin American-Statesman*, March 14, 1989, 12.
58. Jake Pickle, "Challenges for Austin Bergstrom, Environment and the Airport," *Austin American-Statesman*, September 21, 1991, 19.
59. Valarie Bristol, interview by Jennifer Bristol, December 2, 2022.
60. Marla Williams, "Moves to Protect Golden-Cheeked Warbler Have Some Texans Up in Arms, *Corpus Christi Caller-Times*, Sunday July 30, 1995, 17.
61. Ralph K. M. Haurwitz, "House Panel Approves BCCP but Exempts Some Developers," *Austin American-Statesman*, May 19, 1993.

62. "Tax Increment Financing (TIF)," accessed on June 19, 2025, https://en.wikipedia.org/wiki/Tax_increment_financing.
63. Valarie Bristol, interview by Jennifer Bristol, December 2, 2022.
64. "Protests, Fanfare Usher in Balcones Plan," *Austin American-Statesman*, May 3, 1996, 7.
65. Ralph K. M. Haurwitz, "Bill Could Undo Compromise in Balcones Canyonlands," *Austin American-Statesman*, April 3, 1997, 79.
66. Valarie Bristol, interview.
67. Valarie Bristol, interview.
68. Advertisement by Hill Country Alliance, public meeting, *Westlake Picayune*, Thursday October 20, 2005.
69. Valarie Bristol, interview.
70. Charles Bergh, Travis County press release, November 9, 2005.

Chapter 6. The Ladies of the Sea

1. Evelyn Sizemore, *The Turtle Lady: Ila Fox Loetscher of South Padre* (Republic of Texas Press, 2002), 37–42.
2. Tom Longden, "The Turtle Lady Was Also an Aviation Pioneer," *Des Moines Register*, September 23, 2007.
3. Sizemore, *Turtle Lady*, 98–108.
4. Sizemore, *Turtle Lady*, 109.
5. Mary-Love Bigony, "Solving the Ridley Riddle," *Texas Parks and Wildlife Magazine*, March 2004.
6. "Ila Fox Loetscher," Sea Turtle, Inc., accessed 2020, https://seaturtleinc.org/about/ila-loetscher/.
7. George Carmack, "A Day with the Turtle Lady," *San Antonio Express-News*, July 5, 1975, 16.
8. "Woman, 80, Is Florence Nightingale of Turtles," *Longview (TX) News-Journal*, November 4, 1984, 75.
9. "South Padre's Turtle Lady Dies at 95," *Victoria (TX) Advocate*, January 7, 2000, 7.
10. Donna Shaver, interview by Jennifer L. Bristol, February 2022.
11. Shaver, interview.
12. "Environmentalists Threaten Lawsuit over TEDs Decision," *Monitor* (McAllen, TX), July 26, 1989.
13. "Donna J. Shaver, Ph.D.," Samuel J. Heyman Service to America Medals, Partnership for Public Service, accessed 2020, https://servicetoamericamedals.org/honorees/donna-j-shaver-ph-d/.
14. Asher Elbein, "A Texas Program Has Achieved Remarkable Success in Protecting Rare Sea Turtles. So Why Does the National Park Service Want to Defund It?," *Texas Monthly*, August 19, 2020, https://www.texasmonthly.com/news-politics/a

-texas-program-has-achieved-remarkable-success-in-protecting-rare-sea-turtles-so-why-does-the-national-park-service-want-to-defund-it/.

15. "Opinion: Sea Turtle Program Volunteers Stand Behind Donna Shaver, Demand Explanations," *Corpus Christi Caller-Times*, October 28, 2021.
16. Craig Nazor, "What Does it Take to Save a Sea Turtle on Turtle Island?" Siera Club, Lone Star Chapter, accessed December 2023, https://www.sierraclub.org/texas/blog/2023/10/what-does-it-take-save-sea-turtle-turtle-island.
17. Courtney Naquin, "For the Bay, From the Heart," Sierra Club, Lone Star Chapter, April 23, 2021, https://www.sierraclub.org/texas/blog/2021/04/for-bay-heart.
18. Karen McVeigh, "Nurdles: The Worst Toxic Waste You've Probably Never Heard Of," *The Guardian*, November 29, 2021.
19. Anna Berti Suman and Sven Schade, "The Formosa Case: A Step Forward on the Acceptance of Citizen-Collected Evidence in Environmental Litigation?," *Citizen Science: Theory and Practice* 6, no. 1 (2021): 1.
20. Lily Moore-Eissenberg, "No More Nurdles? Formosa's Agreement to Stop Pumping Plastics Into Lavaca Bay is Historic," *Texas Monthly*, October 16, 2019.
21. San Antonio Bay Estuarine Waterkeeper et al. v. Formosa Plastics Corp, Texas et al., No. 6:17-cv-0047 (United States Courts Opinions June 27, 2019).
22. "Overview," Formosa Plastics Corporation Texas, accessed January 9, 2023, https://www.fpctx.com/about-us-main/overview.
23. Delger Erdenesanaa and Lise Olsen, "A Superthreat to the Gulf Coast's Lavaca Bay," *Texas Observer*, February 18, 2022.
24. Erdenesanaa and Olsen, "Superthreat to the Gulf."
25. Diane Wilson, interview by David Todd, Texas Legacy Project, October 23, 2003, https://www.texaslegacy.org/narrator/diane-wilson/.

Chapter 7. The Amplifiers

1. Teresa Tomkins-Walsh, "Houston's Environmental Legacy: Terry Hershey, Community, and Action," *Houston History Magazine*, Spring 2013, 3.
2. Tomkins-Walsh, "Houston's Environmental Legacy," 4.
3. Andrew Sansom, email to Jennifer Bristol, December 13, 2023.
4. Tomkins-Walsh, "Houston's Environmental Legacy," 6–7.
5. Lydia Saldaña, "Force of Nature," *Texas Parks and Wildlife Magazine*, January/February 2015.
6. Janet Warren, "Fate of Big Springs Park to Be Determined Aug. 27," *Abilene (TX) Reporter-News*, August 19, 1992, 2.
7. Rickey Dailey, "Commission Approves Gas Drilling in Falcon State Park," *Monitor* (McAllen, TX), January 22, 1993, 1.
8. Mike Legget, "The Talk of Parks, and Now of Threats," *Austin American-Statesman*, August 31, 1996, 11.

9. Saldaña, "Force of Nature."
10. Andrew Sansom, email to Jennifer Bristol, October 21, 2022.
11. "Terese Tarlton Hershey, 1923–2017," Legacy.com, accessed May 16, 2025, https://www.legacy.com/us/obituaries/houstonchronicle/name/terese-hershey-obituary?id=17297949.
12. "Terese Tarlton Hershey, 1923–2017."
13. George Bristol, conversation with Jennifer Bristol, October 21, 2022.
14. Ann Hamilton, interview by Jennifer Bristol and Valarie Bristol, personal, July 2019.
15. Hamilton, interview by Bristol.
16. Hamilton, interview by Bristol.
17. Ann Hamilton, interview by David Todd, Texas Legacy Project, October 21, 2003, https://www.texaslegacy.org/narrator/ann-hamilton/.
18. Ann Hamilton, "An Analysis of 18 Years at the Houston Endowment: September 3, 1991–October 1, 2009," in author's possession, n.d.
19. George Bristol, interview by Jennifer Bristol, personal, n.d.
20. Ken Kramer, ed., *The Living Waters of Texas*, photographs by Charles Kruvand (Texas A&M University Press, 2010).
21. "Middle Neches Forestland," Conservation Fund, accessed 2021, https://www.conservationfund.org/projects/middle-neches-forestland.
22. "Audubon Texas Celebrates the Legacy of Female Leadership in Texas Conservation Efforts," Audubon Texas, February 9, 2016, https://tx.audubon.org/press-release/audubon-texas-celebrates-legacy-female-leadership-texas-conservation-efforts-0.
23. Ann Hamilton, email to Jennifer Bristol, December 20, 2021.
24. Hamilton, interview by Bristol.
25. Darlene Lewis, interview by Jennifer Bristol, personal, August 2019.
26. "Darlene Lewis Joins KOKE Music Station," *Austin American-Statesman*, March 20, 1985, 42.
27. Lewis, interview.
28. Lewis, interview.
29. Lewis, interview.
30. "CO-OP Adds 20,000 to BCPD Activity Fund," *Facts* (Clute, TX), February 19, 1998, 13.
31. Lewis, interview.
32. Data provided by Carley Blankenship, Community Outdoor Outreach Program grants manager, Texas Parks and Wildlife Department.
33. Marty Leonard, interview by Jennifer Bristol, 2019.
34. Leonard, interview.
35. "Marty Leonard Community Chapel Receives Architecture Award," *Fort Worth Business Press*, January 25, 2016.

36. Lana Gilson, "Lena Pope Early Learning Center," Nature Explore Program, accessed May 16, 2025, https://certified.natureexplore.org/lena-pope-early-learning-center.
37. "Protecting What We Love," Friends of the Fort Worth Nature Center, accessed August 27, 2022, https://naturecenterfriends.org/.
38. "Vote: Water Board Election Could Be Referendum on Trinity Uptown Plan," *Fort Worth Star-Telegram*, April 30, 2006, 32.
39. *Marty Leonard Tribute: Fort Worth Wild Award* (video). (Friends of the Fort Worth Nature Center and Refuge, 2021).
40. "Water Supply," Tarrant Regional Water District, accessed May 16, 2025, https://www.trwd.com/water-supply/.
41. "Water Supply."
42. "Environmental Sustainability," Tarrant Regional Water District, accessed May 16, 2025, https://www.trwd.com/environmental/.
43. Leonard, interview.
44. Sandra Sobieraj Westfall, "Laura Bush Shares Her Camping Photo Album—and Says Roughing It Is More Fun with Girlfriends Than George," *People*, December 1, 2020.
45. Michelle Brownstone, "Laura Bush Hopes to Inspire Conservation," *Odessa (TX) American*, March 2, 2013, A1.
46. "Native Texas Park," George W. Bush Library, accessed August 26, 2022, https://www.georgewbushlibrary.gov/explore/native-texas-park.
47. "A Legacy of Conservation—a Conversation with Regan Gammon," Texan by Nature, accessed May 16, 2025, https://texanbynature.org/2019/09/conversation-with-regan-gammon/#.
48. "Conservation Wrangler," Texan by Nature, accessed May 16, 2025, http://texanbynature.org/programs/conservation-wrangler/.
49. "Coalition Launches Texas Natural Resource/Environmental Literacy Plan," Texas Parks and Wildlife Department, January 22, 2013, https://tpwd.texas.gov/newsmedia/releases/?req=20130122c.
50. Brownstone, "Laura Bush Hopes."
51. Jennifer L. Bristol, "Monarch Butterfly—a Blog I Wrote Ten Years Ago," Atomic Cowgirl, accessed August 26, 2022, https://theatomiccowgirl.com/cowgirl-notes/monarch-butterfly-a-blog-i-wrote-ten-years-ago.
52. "Laura Bush, Daughter Writing Children's Book," *Longview (TX) News-Journal*, October 30, 2015, A2.

Chapter 8. The Land Trusts

1. Carolyn Vogel, interview by Jennifer Bristol and Valarie Bristol, personal, September 27, 2022.
2. Vogel, interview.

3. Vogel, interview.
4. "The State Park Board," *Waxahachie Daily Light* (Ellis County, TX), May 14, 1946, 4.
5. Vogel, interview.
6. Vogel, interview.
7. Vogel, interview.
8. Vogel, interview.
9. Vogel, interview.
10. Vogel, interview.
11. "About," Texas Land Trust Council, accessed September 21, 2022, https://texaslandtrustcouncil.org/about/.
12. Rosemary Barnes, "Bay Group Plans New Strategy," *Corpus Christi Caller-Times*, January 19, 1998, 13.
13. Lynn Brezonsky, "Land Trusts Put Dent in Urban Sprawl," *Abilene (TX) Reporter-News*, January 1, 2004, 12.
14. Diane Jennings, "Keeping the Prairie Sky Wide and High," *Corpus Christi Caller-Times*, September 1, 2002, 17.
15. Carter Smith, letter of support to Audubon Texas Women in Conservation Terry Hershey Award, 2014, Carolyn Vogel personal correspondence.
16. Vogel, interview.
17. Vogel, interview.
18. Vogel, interview.
19. Siglo Group, *Valuing Economic Benefits of Texas Conservation Lands* (Texas Land Trust Council, 2019), https://texaslandtrustcouncil.org/wp-content/uploads/2019/03/TLTC_EcosystemServiceValuation_Report_wAppendix_2019_03_01_Final.pdf.
20. Texas Land Trust Council, home page, accessed September 21, 2022, https://texaslandtrustcouncil.org/.
21. Kassi Scheffer-Geeslin, email to Jennifer Bristol, January 24, 2023.
22. "From Harvard to Houston," *Evening Express* (Portland, ME), January 2, 1985, 2.
23. Mary Anne Piacentini, letter of recommendation for Carolyn Vogel, 2014. Personal correspondence.
24. Marty Schladen, "Highway Plan at Crossroads in the County," *Galveston Daily News*, August 15, 1999.
25. Jaime Gonzalez, interview by Jennifer Bristol, October 9, 2022.
26. Mary Ann Piacentini, interview by Jennifer Bristol, October 14, 2022.
27. Gonzalez, interview.
28. Piacentini, interview.
29. "Mary Anne Piacentini," Coastal Prairie Conservancy, accessed October 27, 2022, https://www.coastalprairieconservancy.org/mary-anne-piacentini.
30. Blair Fitzsimons, interview by Jennifer Bristol, November 7, 2022.
31. Fitzsimons, interview.
32. Fitzsimons, interview.

33. Fitzsimons, interview.
34. No Land, No Water, home page, accessed November 7, 2022, https://www.nolandnowater.org/.
35. Texas Agricultural Land Trust, *2020 Annual Report* (Texas Agricultural Land Trust, 2020), n.p.
36. Jena McRell, "For Love of the Land," *Cattleman*, December 2022.
37. Blair Calvert Fitzsimons, "Bill Gives Incentive to Save Land from Development," *Austin American-Statesman*, July 15, 2014, A8.
38. Texas Agricultural Land Trust, home page video, "On Mission . . . Together," accessed June 22, 2025.
39. No Land, No Water, home page.
40. Blair Calvert Fitzsimons, "Act Today to Protect and Preserve Rural Texas Lands," *San Antonio Express-News*, February 12, 2017.
41. "Impact," Texas Agricultural Land Trust, accessed November 8, 2022, https://www.txaglandtrust.org/impact/.
42. "Outdoorsman of the Year Award, Blari Fitzsimonds," posted October 17, 2024, by Texas Wildlife Association Foundation, YouTube, 4 min., 22 sec., https://www.youtube.com/watch?v=NLAGCa_1W74.
43. "Calvert Brothers Ranch," Texas Agricultural Land Trust, accessed November 8, 2022, https://www.txaglandtrust.org/portfolio/calvert-brothers-ranch/.
44. Blair Fitzsimons, LinkedIn profile, accessed November 6, 2022, https://www.linkedin.com/in/blair-fitzsimons-40b4682a.

Chapter 9. The Storytellers

1. Karle Wilson Baker, *The Birds of Tanglewood* (Southwest Press, 1930).
2. Sarah Ragland Jackson, *Texas Woman of Letters: Karle Wilson Baker* (Texas A&M University Press, 2005), 8–10.
3. Martha Anne Turner, *Old Nacogdoches in the Jazz Age* (Madrona, 1976), 46
4. "Karle Wilson Baker," Stephen F. Austin State University, accessed 2021, http://www.sfasu.edu/story/markers/Baker-marker-dedication.html.
5. Edwin W. Gaston Jr., "Baker, Karle Wilson (1878–1960)," Texas State Historical Association, 1976, last updated August 3, 2017, https://www.tshaonline.org/handbook/entries/baker-karle-wilson.
6. Lydia Saldaña, interview by Jennifer L. Bristol, personal, January 10, 2020.
7. Saldaña, interview.
8. Saldaña, interview.
9. *Nature Tourism in the Lone Star State: Economic Opportunities in Nature: A Report from the State Task Force on Texas Nature Tourism* (Texas Parks and Wildlife Department and Texas Department of Commerce, n.d.).
10. Darcy Bontempo, interview by Jennifer L. Bristol, personal, May 19, 2022.

11. "'Life's Better Outside' for Texas Young People, Families," press release, Texas Parks and Wildlife Department, May 22, 2006.
12. "The State of Water," Texas Parks and Wildlife Department, accessed 2022, http://www.texasthestateofwater.org/.
13. Saldaña, interview.
14. Lydia Saldaña, résumé, shared with Jennifer L. Bristol, 2022.

Chapter 10. The Collaborators

1. Thea Platz, interview by Jennifer Bristol and Valarie Bristol, personal, September 30, 2022.
2. William Pack, "Master Naturalist Volunteer Group Pushes Public Service," *San Antonio Express-News*, November 5, 2000, 7B.
3. Platz, interview.
4. Christopher Anderson, "Park, Naturalist Group, SAWS to Be Honored for Environmental Work," *San Antonio Express-News*, May 1, 2001, 7B.
5. "Walker Ranch Historic Landmark Park," City of San Antonio, accessed May 19, 2025, https://www.sa.gov/Directory/Departments/Parks/Parks-Facilities/Parks/Directory/Walker-Ranch-Historic-Landmark-Park.
6. Platz, interview.
7. Platz, interview.
8. Platz, interview.
9. Platz, interview.
10. Platz, interview.
11. "About," Texas Master Naturalist Program, accessed October 9, 2022, https://txmn.tamu.edu/about/.
12. Nancy Herron, interview by Jennifer Bristol, August 2019.
13. Herron, interview.
14. Richard Louv, *Last Child in the Woods: Saving Our Children from Nature-Deficit Disorder* (Algonquin Books of Chapel Hill, 2006).
15. Hayden Brooks, interview by Jennifer Bristol, 2019.
16. Molly Stevens, interview by Jennifer Bristol, September 2019.
17. Herron, interview.
18. Eliot Shapleigh and Donna Howard, Texas Partnership for Children in Nature, 81(R) History for SB 205, Texas Legislature Online, May 2009, https://capitol.texas.gov/billlookup/History.aspx?LegSess=81R&Bill=SB205.
19. Texas Parks and Wildlife Department, *Texas Children in Nature Strategic Plan* (Texas Parks and Wildlife Department, 2010).
20. Texas Parks and Wildlife Department, *Texas Children in Nature*.
21. Brooks, interview.

Chapter 11. The Private Landowners

1. Pat Shield Ayres and Vera Ayres Bowen, interview by Jennifer Bristol and Valarie Bristol, personal, 2021.
2. "Woman's Body Found in Creek," *Austin American-Statesman*, June 19, 1962, 1.
3. "TYC Develops Bill of Rights," *Marshall (TX) News Messenger*, December 5, 1975, 8.
4. "Texas Youth Council Head Stripped of Voting Rights," *Marshall (TX) News Messenger*, December 5, 1975, 8.
5. "TYC Serves Youth Needs," *Marshall (TX) News Messenger*, December 5, 1975, 8.
6. Ayres and Bowen, interview.
7. *Shield Ranch Master Plan* (Shield Ranch, 2018), 12.
8. *Shield Ranch Master Plan*, 29.
9. Dick Stanley, "Ranch Owners Lauded for Preserving Texas Habitats," *Austin American-Statesman*, May 31, 2000, 65.
10. Ayres and Bowen, interview.
11. Shield Ranch, *Camp El Ranchito Capital Campaign*, brochure (Shield Ranch, 2019).
12. Justine Kies, email to Jennifer Bristol, October 20, 2022.
13. Shield Ranch, *Camp El Ranchito Capital Campaign*, brochure (Shield Ranch, 2019).
14. "Can't End Charges: Students Stalk Out of Erwin Session," *Fort Worth Star-Telegram*, October 24, 1969, 11.
15. "Frank Craig Erwin Jr.," Texas State Cemetery, accessed May 20, 2025, https://cemetery.texas.gov/locate-a-plot/plotholder/frank-craig-erwin-jr.
16. Paula Smith, interview by Jennifer Bristol and Valarie Bristol, personal, October 13, 2022.
17. Kenneth Bunting, "Bush's Son Gets Money Tip from Office of Possible Rival," *Fort Worth Star-Telegram*, April 22, 1989.
18. "Stocks," *Fort Worth Star-Telegram*, January 31, 1990, 50.
19. Paula Smith, interview.
20. Paula Smith, interview.
21. Margie Crisp, *The Nueces River: Río Escondido*, with artwork by William B. Montgomery (Texas A&M University Press, 2017), 71
22. Paula Smith, interview.
23. Paula Smith, interview.
24. Paula Smith, interview.
25. Carter P. Smith, "Deer Hunt on the Dobbs," *Texas Parks and Wildlife Magazine*, October 2013.
26. Paula Smith, interview.
27. Carter Smith, email to Jennifer Bristol, January 4, 2023.
28. Carolyn Ohl-Johnson, "A Little Christmas Mountains History," *Christmas Mountains Oasis* (blog), December 1, 2022, https://cmoasis.blogspot.com/2022/12/a-little-christmas-mountains-history.html.

29. "Buckskinners Rendezvous and Indian Pow Wow," *Jeff Davis County News*, September 28, 1988, 1.
30. Gary Clark, "Christmas Mountains Oasis a Haven for Birds," *Houston Chronicle*, August 26, 2011.
31. Cecilia Nasti, "West Texas Wetland," *Texas Parks and Wildlife Passport to Texas*, aired August 1, 2018.
32. Carolyn Ohl-Johnson, interview by Jennifer Bristol, April 16, 2022.

Chapter 12. The Future

1. Daniel J. Middleton, "Danielle Belleny, Wildlife Biologist and Co-Organizer of #BlackBirdersWeek," *Unique Coloring* (blog), accessed May 20, 2025, https://www.uniquecoloring.com/articles/danielle-belleny.
2. Danielle Belleny, interview by Jennifer Bristol, 2021.
3. Black AF in STEM, home page, accessed September 22, 2022, https://www.blackafinstem.com/.
4. Virginia Rose, interview, *BirdCallsRadio*, aired July 10, 2018.
5. "From Our Founder," in *Birdability 2021 Annual Report* (Birdability, 2022), 2.
6. "Press and Publicity," Birdability, accessed November 5, 2022, https://www.birdability.org/press-and-publicity.
7. "From Our Founder," 2.
8. Marisa Oliva-Rodriguez, interview by Jennifer Bristol, 2020.
9. Oliva-Rodriguez, interview.
10. Bob Powell, "Local Woman Lives Community Involvement Through Work," *Monitor* (McAllen, TX), March 26, 2006, A66.
11. Roy Rodriguez, "Kids Club Is Strictly for the Birds," *Monitor* (McAllen, TX), October 23, 2005, 73.
12. *Dream Big: Living Your Bucket List*, XXIV Summit Guide (Association of Nature Center Administrators, 2018), https://natctr.org/images/pdfs-doc/summit/2018_ANCA_Guide.pdf.
13. Oliva-Rodriguez, interview.
14. Elene Bruess, "After over Century, First Herd of Buffalo Return to Native Lands of Lipan Apache," *San Antonio Express-News*, December 6, 2021.
15. Dylan Baddour, "Coming Home," *Nature Conservancy Magazine*, November 10, 2022, https://www.nature.org/en-us/magazine/magazine-articles/texas-buffalo-project/.
16. Baddour, "Coming Home."
17. Texas Tribal Buffalo Project, *2022 Annual Report* (Texas Tribal Buffalo Project, 2022).
18. Baddour, "Coming Home."
19. Bruess, "After over Century."

20. "About Dr. Katharine Hayhoe, Climate Scientist and Activist," Katharine Hayhoe's personal website, accessed May 20, 2025, http://www.katharinehayhoe.com/biography/.
21. "Katharine Hayhoe, Ph.D.," Department of Political Science, Texas Tech University, accessed May 26, 2022, https://www.depts.ttu.edu/politicalscience/faculty/hayhoe_katharine.php.
22. "Sisters on the Planet," Oxfam, accessed May 20, 2025, https://www.oxfamamerica.org/about-us/ambassadors/sisters-on-the-planet/.
23. "Katharine Hayhoe," Nature Conservancy, accessed May 26, 2022, https://www.nature.org/en-us/about-us/who-we-are/our-people/katharine-hayhoe/.
24. Sonia Smith, "Unfriendly Climate," *Texas Monthly*, May 2016.
25. Gina Hinojosa, "A National Climate Plan Needs a Texas Component," *Austin American-Statesman*, April 8, 2021.
26. "About Dr. Katharine Hayhoe."

INDEX

Pages in italics indicate photographs.

ABOUT THE AUTHOR

Jennifer L. Bristol is the author of *Parking Lot Birding: A Fun Guide to Discovering Birds in Texas* and *Cemetery Birding: An Unexpected Guide to Discovering Birds in Texas*. She contributed to her father's book, *Texas State Parks: The First 100 Years, 1923–2023*. Sharing a love of birds and nature with famed naturalist Victor Emanuel, she created and directs the Victor Emanuel Nature Scholarship Fund to send underserved youth to birding and nature immersion camps. Having a deep sense of service, she has served as chair of the Environmental Commission, vice chair of the Downtown Commission for the City of Austin, vice chair of the Travis County Bond Citizen Advisory Board, and on numerous national and local boards. Bristol resides in Austin with her husband, five charismatic dogs, and three horses.